Advance Praise for Career Guide for the High-Tech Professional

"Thanks for doing all job seekers a great service by writing a book that truly tells it like it is! I regard *Career Guide for High-Tech Professionals* as the most accurate and effective guide to self-marketing available. Your book clearly has much to teach even highly sophisticated marketers about how to market themselves."

—Stephen H. Lahey,
Lahey Consulting, LLC

"This all-encompassing guide for knowledge worker employment is remarkable! A perfect blend of theory and practice. It's informative, thorough, and, most importantly, it gives realistic, practical philosophy and steps to take in creating fulfilling and productive economic marriages."

—Dr. Stephen R. Covey, author
The 7 Habits of Highly Effective People

"The essence of this book in two words: Be relevant. No one gives you a job. You earn one, and Perry shows you how."

—Seth Godin, author
Survival is Not Enough

"This is 'guerrilla headhunting' at its best, and I highly recommend that you invest the time in reading it—it could change your life!"

—Jay Levinson, author
Guerrilla Marketing

"*Career Guide for the High-Tech Professional* is a 'playbook' for winners. In this leadership-hungry environment, Perry shows you how to embrace your talents, articulate your worth, and excel in a career you're passionate about."

—Guy Kawasaki, author, *Rules for Revolutionaries*
chief executive officer, Garage Technology Ventures

"It's rare that you read a book where the information is excellent and the style of writing is a pleasure to read. This book is packed with excellent ideas and insights presented in a very readable way. David Perry has written a must-have book for anyone who wants to achieve their career aspirations."

—Lee Silber, author, *Self-Promotion For The Creative Person*
Winner of the 2002 Theodore S. Geisel Award (Best Business Book of the Year)

"Perry's done it again! *Career Guide for the High-Tech Professional* is loaded with valuable insights, information, and practical actionable advice."

—John Reid, president
CATA*Alliance*

"You've managed to write one of the best job search books I've read…period! Most books contain the same old, high-level strategic advice, which has its place, but what I loved about yours was that it got into the nitty-gritty tactical…how to actually execute what other books just advise. Great advice, great strategies, great tips. Great job."

—Ross Macpherson, president
Career Quest

"Next to marriage, your choice of a company and a career can be the most important decisions of your life. This powerful, practical book shows you how to make the right decision in a fast changing market!"

—Brian Tracy, author
Focal Point

"David Perry has applied new economy digital smarts to the now economy's tougher reality and thrown in an old economy dose of common sense. You can sit there and wait for a job to happen to you, or you can read his book and get yourself positioned for a lifetime of challenge and reward."

—Nathan Rudyk, vice president of marketing,
Databeacon Inc.

"Perry is the most creative and successful headhunter of the bunch. From the perspective of someone who knows what to look for in a candidate he spills the beans on how to make yourself look like what he would hire. It's like getting the combination to the vault straight from Scrooge. Take the lesson and apply."

—Tony Patterson, publisher (retired)
Silicon Valley North Newspaper

"I just finished *Career Guide for the High-Tech Professional* and want to thank you for the experience. I found it totally delightful, filled with current, down-to-earth advice and guidance. Thanks for giving me the respect of teaching and suggesting instead of forcing me to adopt another 'if you do this, salvation will come' methodology. I was also quite impressed with your Website, especially with the ease and user-friendliness of the paper-free distribution for your book."

—Pavla Selepova, founder and CRM guru,
MiroMetrica

"Hello David, Kudos on your book. I read it once again and feel it very accurately reflects the way the recruiting cycle works. By the way, this recruiter is okay with people reading your book. The simple fact, as you state in the book, is that most individuals are unwilling to follow this very sound approach to job hunting. As a result, people find positions through chance, mischance, and to satisfy others' needs. If more individuals identified their passion, researched the market to identify companies that could advance it, and pursue those companies by research and planning, there would be far fewer unhappy employees commiserating with Dilbert."

—Jack O'Brien
Hire Integrity, Elgin, IL

CAREER GUIDE

FOR THE

HIGH-TECH

PROFESSIONAL

Where the Jobs Are Now and How to Land Them

David Perry

Foreword by Jay Conrad Levinson,
author of *Guerrilla Marketing*

CAREER
PRESS

THE CAREER PRESS, INC.
Franklin Lakes, NJ

CAREER GUIDE FOR THE HIGH-TECH PROFESSIONAL
EDITED BY JODI BRANDON
TYPESET BY STACEY A. FARKAS
Cover design by DesignConcept
Printed in the U.S.A. by Book-mart Press

To order this title, please call toll-free 1-800-CAREER-1 (NJ and Canada: 201-848-0310) to order using VISA or MasterCard, or for further information on books from Career Press.

The Career Press, Inc., 3 Tice Road, PO Box 687,
Franklin Lakes, NJ 07417
www.careerpress.com

Library of Congress Cataloging-in-Publication Data

Perry, David, 1960 Jan. 12-
    Career guide for the high-tech professional : where the jobs are now and how to land them
/ by David Perry ; foreword by Jay Conrad Levinson.
        p. cm.
    Includes index.
    ISBN 1-56414-743-6 (paper)
    1. High technology industries—Vocational guidance—United States. 2. Job
hunting—United States. 3. Vocational guidance—United States. I. Title.

HF5382.5.U5P39 2004
650.14--dc22

                                                                                    2003069599

This book is dedicated to my five miracles:
My wife and business partner, Anita Martel, the love of my life; and our four amazing
children, who don't know the meaning of "not possible." You all inspire me.
Christa (the actress/detective)
Corey (the entrepreneur/tech-wizard)
Mandy (the artist/singer)
Shannon (the dancer/fashion model)

To my parents, Myrtle Gallagher and Fred Perry, who have given more of themselves
than any parents should. You are both incredible. Thanks.

Acknowledgments

This book never would have come into being if not for three key people in my life:
Bob Henault, who saw in me the "seeds of a great recruiter"; Jerry Alary, who
challenged me to write the book; and Ron Wiens, who assured me that I did know
enough about this "career business" to actually help others. This book is, in a sense,
entirely your fault.

My clients, many of whom have become close friends, are mavericks, evangelists,
and pioneers. I was astounded by how you pitched in to read drafts, make
suggestions, and keep the book relevant and up to the moment. Thank you so much
to Peter Kemball, Tony Patterson, John Reid, Pavla Selepova, Bob Cook, Paul
Swinwood, Rick Crutchlow, Kevin Watson, Charles Duffet, Simona Featherling, Anne
and Bob Valcov, Nathan Rudyk, Kate Beere, Andrew Beere, Mark Taylor, Dieter
Hensler, Chris Fedorko, Bill Breen, Phil Read, Jim Reil, Gordon Deans, Gail (Rainbow)
Merrikin, Shari Miller, Jean-Michael Voltan, Daniel Houle, Josh Korn, George Pytlik,
Barry Gander, Brian Clark, Blake Carruthers, Steve Waldo, Lynda Partner, Arthur Young,
Allan Pace, Brenda Batton, Nigel Parker, Gerry Turcotte, Jean-Pierre Jauvain, Jeff Potts,
Corein Kershey and many others who won't openly admit they even know me.

I would also like to thank Nicole Leblanc for typing and retyping more than 700
pages of notes, and Christine Bertrand, who kept me focused on our clients's projects
during the daylight hours. Sequel anyone?

Thanks to everyone at Career Press, especially, Stacey, Kirsten, Jodi, and Mike.

Also, I need to underline the positive impact Jay Conrad Levinson has had on my
career. Jay went out of his way to help me get this book into your hands. Jay, without
your business manuals I would never have made it this far. On behalf of the millions of
readers just like me, thanks you for writing *all* of the *Guerrilla Marketing* books.

Lastly I must thank my father, Fred Perry, who endured my prose and terrible
spelling through some nine revisions. I know I owe you another box of red pens, Dad.
I had that there, their, and they're thing down pat, but now I don't no where my notes
are.

Contents

Foreword

by Jay Conrad Levinson

Getting the exact job you want with the company you want is either highly improbable or a cinch. It's highly improbable if you play by the old rules and go about your task in the old-fashioned way. But it can be cinch if you play by the new rules and go about your task in the ways suggested by David Perry in *Career Guide for the High-Tech Professional*.

He has written this book from the inside out. That means he's a professional headhunter who knows the intricacies, the secrets, and the overlooked tactics when it comes to finding your ideal job. If you're willing to look upon the task of finding that job as a job unto itself, you're going to reap exceptional rewards from this book.

It does not contain shortcuts, tricks, or anything unethical. But it does contain ultra-solid advice along with the rich details of landing the job of a lifetime.

David says you don't have to read the entire book in order to benefit from it. He's telling the truth. But if you're honestly motivated to take complete control of your life and your future, you'll read it from cover to cover. In my opinion, not one word on one page has been wasted. And the more you glean from his information, the better job you'll land. It's that simple.

This is not a book about the theory of securing employment. Instead, it's a book filled with job-finding wisdom and the actual things you must say in your cover letter, on your resume, and during your interviews. Nothing is left to chance. Every detail is covered.

Although this book appears to be about getting a job, it's really about living a fulfilling life—because that's the end result of obtaining your ideal job. Few people really understand that. They see a job as a way to earn money rather than a way to thrive and prosper while living the life they've always dreamed of living. The U.S. Department of Commerce tells us that 80 percent of employed people are not really content with their jobs.

Readers of this book—that is, readers who follow the advice it offers—will be in the 20 percent of people who wouldn't dream of leaving their job. That can be you—but only if you take action based upon what you'll learn once you've read this insightful book.

Although David Perry has achieved remarkable success as an executive recruiter, I believe that he wouldn't have achieved as much if a book such as this existed before he began his career. Reading it, you'll learn as much as or more than many current executive recruiters know. Acting upon it, you'll surpass your most ambitious fantasies. For the real truth is that landing the job of your dreams is not as difficult as you may think. It's all a matter of being your own headhunter, and that's exactly what you'll learn in the pages ahead.

You'll learn how to navigate masterfully in a new economy, in a world more high-tech than it has ever been before. You'll learn to see things from the standpoint of the people who will hire you rather than from your own perspective.

Even the title of the book gives you a clue as to what it offers. There's an old Chinese proverb that says, "If you have foresight, you're blessed, but if you have insight, you're a thousand times blessed."

This is a book of insights. It's a book about you—your future, your sense of well-being, your life. Getting the kind of job that can change your life for the good—forever—isn't a matter of being on top of most of the details. It's a matter of mastering *all* the details. And that's what you're about to learn.

I welcome you to this book, these insights, this information, and most of all, to the life you've dreamed of living. This is the kind of book that makes those dreams come true.

—Jay Conrad Levinson
author, *Guerrilla Marketing* book series

How a Job Gets Filled
Today's New Reality and the
Lessons to Be Learned

How a Job Gets Filled—From the Employer's Point of View

Week 1

Monday

- 8 a.m.: It's early Monday morning and John Smith the development manager for the O-So-Big Software company, is barely into his third cup of coffee and he's already having a bad day—a potentially fatal one for his career. One of his star programmers just resigned. John tried to talk him out of it to no avail. Even though John was angry he was smart enough to offer Bill Guy the programmer a 20-percent raise. John figured that would keep Bill with the company just long enough for him to hire someone to replace him and then he would fire Bill. No one leaves on his watch. John is on a tight deadline, which he can't miss. Now John wonders who he can promote. No one.

- 12:00 p.m.: By noon John has called everyone in the department to ask if they know anyone qualified to do Bill's job.

  > As much as 80 percent of all hiring happens at this juncture. Candidates are chosen from a pool of friends and colleagues suggested by employees. People hire their friends because they are known quantities and not shots-in-the-dark! The more senior the role, the more that rule applies.

- 3:00 p.m.: John calls all his friends at other companies to see whom they know. No luck. He then checks his favorite newsgroups, his most trusted sources, for leads they might have. Newsgroups are a great source of like-minded people who share information, even job leads, with each other. No luck.
- 5:00 p.m.: John, per his company's personnel policy, requests resumes from HR. They'll get back to him tomorrow.

Tuesday
- 12:00 p.m.: HR has dropped off a box of resumes for his review. It is a very large box full of yellowed resumes.
- 2:00 p.m.: John asks his executive assistant, Jill, to look through the box and find a "team leader with 5 years C ++ and JAVA development." Jill looks for the key words C++, JAVA, and team leader. If a resume is missing any of the three it's discarded. Jill sifts through a few hundred resumes in an hour while answering the phone and preparing a report for the vice president of engineering.

> *The initial resume screening is often times delegated to a junior staff member who's been instructed to sort through the resumes looking for three to five key words. If your resume contains those words, you will be screened in. If not, you're out.*

Thursday
- 10:00 a.m.: Jill gives John the resumes of three possible candidates. Hurray! He will start interviewing Monday of next week.

> *As many as 50 percent of candidates are hired from "old" resumes kept on file. This explains why HR managers run ads during times of economic slow-downs even when they don't have any openings. They're collecting resumes to be used later.*

Week 2

Monday
- John meets Candidate 1. Too expensive.

Tuesday
- John meets Candidate 2. Too little experience.

Wednesday
- John meets Candidate 3. Doesn't really qualify.

Thursday
- No one is really strong enough, so John asks HR to recruit more. Now he's really under pressure.

Friday
- 3:00 p.m.: HR sends the paperwork for John to fill out over the weekend so they can run an ad in the local newspapers next week and post it on a few job boards.

Week 3

Friday

◆ After several iterations of the "copy," the wording for the newspaper ad is approved.

Week 4

Monday

◆ Resumes come in. HR reviews.

> *You have a 5- to10-percent chance of landing a job through a newspaper ad when the economy's booming. This goes down to less than 1 percent during a recession.*

Tuesday

◆ Headhunters start calling. John loses most of his day talking to headhunters. Many of them sound as though they have good candidates. He really needs to get going right now, but HR spent the money on the ad and wants to see the results first. Besides, John says, "Why pay a fee if I don't have to?"

Wednesday

◆ John gives up trying to be polite and says no to the hundreds of headhunters who've phoned trying to get a job order. He's decided to let a select few (10 to 20) to look for him on a contingency basis. John figures, "What's the harm? Anyone actively looking will respond to my ad." Contingency headhunters only get paid if one of their candidates gets hired. "Their candidate will have to be vastly superior to the ones I get before I'll pay a fee," remarks John.

> *You have a 1-percent chance of successfully landing a job with a contingency recruiter during boom times. This number drops during a recession.*
>
> *On the other hand, you have a 10 percent chance of landing a job through a retained executive search firm, and the odds go up during a recession because the recruiter has been hired to specifically "headhunt" a gainfully employed individual.*

Week 5

Monday

◆ Resumes are still coming in. HR is still reviewing them. Headhunters' submissions are put aside for later review.

Week 6

◆ HR does vital screening interviews for John to assess corporate fit.

Week 7

Tuesday

◆ John meets a great candidate who applied through an ad. He's about to extend an offer when he learns that two competing headhunters have also presented his number-one candidate, who answered his ad. If he hires him, he has to pay two $20K fees. The candidate didn't know the agency had submitted his resume, but it was too late. The damage is done.

Wednesday

◆ John starts interviewing again.

Week 8

◆ John and Jill are interviewing. Headhunters are still calling, and John's project is way behind.

As a Job Seeker, What Should You Learn From This Very Typical Example?

Well, as does every other hiring manager in every other firm, John just wants to hire competent people when he needs them. An astute manager constantly assesses the strengths and weaknesses of his team and spends a certain amount of time "talent scouting" and interviewing people in anticipation of an opening—something John Smith did not do. Those are facts.

As a job seeker, you might have landed a job with John on the Monday of Week 1 of John's fiasco if:

◆ You had contacted John with a memorable resume in the preceding four weeks.

◆ Your network of friends had known about the job *or* knew you were looking.

◆ Your resume answering the ad had stood out from the crowd.

◆ You had a particularly persuasive headhunter working on your behalf who had set up an information interview for you prior to John's need.

Everyone has an opinion on the best way to find a job, and most of the books written today on how to find a job follow one or more of these methods. In today's dot-com bust only one of these traditional methods will work, and that method needs updating.

It's important to understand how you have been traditionally taught to look for a job and why those methods don't work well now so that you can focus all your energy on what does work. The following adaptation from the world-famous book *What Color Is Your Parachute* (Ten Speed Press, 2004), demonstrates the best and worst ways to find a new job according to author Richard Bolles.

Richard definitely knows what he's talking about because *Parachute* has now sold 10 million copies. As you'll see in the following chart, obviously networking and approaching employers in a targeted manner reaps the best results, but with too many people using the same tactics you need to be even more strategic. How to raise your voice above the roar of the crowd and be heard and hired is what you'll learn in this book with step-by-step instructions.

10 Ways to Find a Job According to *Parachute* (Pre-2001)

	Best and Worst Ways to Find a Job, Based on *What Color Is Your Parachute?*	Percentage Chance of Landing Job
The 5 *Worst* Ways To Find A Job	Randomly mail out resumes to employers: You'll be successful every 1 out of 1,700 resumes sent. That's a lot of trees!	7%
	Answer ads in professional or trade journals appropriate to your field: Targeted, yes, but the lead times for publication often mean the job is filled before you see it.	7%
	Answer ads in newspapers in other parts of the state or country: Employers always hire locally first to save on moving costs.	10%
	Answer ads in local papers: Great when there is a boom. Hit-and-miss otherwise.	5–24%
	Use headhunters: This really depends on the salary level and how scarce your skill set is.	5–24%
The 5 *Best* Ways To Find A Job	Personal referrals: Ask family members, friends, people in the community, and staff at career centers, "Do you know of any jobs in my field?"	33%
	Knock on the doors: Contact every company that interests you after you check it out on the Web.	47%
	Yellow Pages: Consult your local Yellow Pages directory to identify companies in your area that you may have an interest in and call them. (Of course, nowadays, I'd use Google first.)	69%
	Job Hunter's Club: Develop your own support group or "Job Hunters Club" and implement the previously mentioned way. I did that 15 years ago and eventually it led me to this field	84%
	Target Marketing: Find a company with whcih you want to work. Find out who has the authority to hire you. Find out what types of problems he or she is facing. Figure out how your skill set can help solve the problems. This is exactly what headhunters do on your behalf when they market you.	86%

A True Story

Let me tell you the story of Lance Cruiser, which illustrates how the market has changed. (Yes, I changed his name). Lance is a top programmer with ABC Widgetware. He's always had stellar reviews and has been asked on numerous occasions to join high-profile teams working on bleeding-edge designs with his employer. By all accounts Lance is one hot commodity: You only need to listen in on some of the calls from headhunters and friends who have tried to lure him away over the years.

Unfortunately one morning Lance's bleeding-edge, next-generation project just got axed! It seems the company won't see any revenue from it for two years and that's no longer a luxury the company can afford. As with many other companies these days, it needs revenue now or there will be bigger cuts to follow. Sound familiar?

Lance accepts a nice severance package, figuring that getting another gig would be a breeze. He knows many of his friends had been looking for jobs without much luck, but Lance is different. People are always calling him for advice. Now it's time for him to call in a few favors and look at some new opportunities.

For Lance, days turn into weeks and weeks into months as he looks for a new job using the table on page 15, the same way he always used to. Over the course of the first seven months of his job search he:

- Writes to more than 150 companies that had advertised jobs in newspapers, sending professionally prepared letters and a professionally designed CV to all. To these, he receives no response at all.

- Registers with several hundred job boards through a service he subscribes to, and personally takes the time to download his resume to the top five job boards. After two months, only two unknowns have retrieved his CV.

- Sends more than 430 e-mails.

- Places well more than 1,000 calls.

The results are abysmal.

Finally, Lance decided the old ways aren't working. He is frustrated and depressed. Out of options, he decides to use the last way indicated on the table: Target Marketing. He will approach employers directly. Thinking about that, Lance has to sit back and analyze what the message needs to be for each employer. After deciding on the best value proposition, (refer to The New Value Table in Chapter 1), Lance analyzes the local market and starts his campaign by making the following assumptions: He will do all the heavy lifting. Headhunters, job boards, Websites, and friends can't guarantee results; he has a greater vested interest. In fact, he has the *only* vested interest. He doesn't expect to find work through the newspapers or job search Websites. Most would use these mediums as the primary lead source. It's the path of least resistance. As he says, "Every dog is sniffing at the same gopher hole. And it's tough to separate yourself from the pack when you're leashed to it." Also, the best jobs are rarely advertised.

Lance soon realized that it takes more effort and ability to target specific companies, find the contact, get attention, and develop a position. But he also discovered that fewer people do this precisely because of the work involved. Tertiary benefits are the relationships developed in advance of requirements. Ironically, it's much easier to position yourself for a job when the job doesn't exist.

In the end, Lance identified two companies with significant problems he could solve. He wrote a personal professional letter to both companies's senior vice presidents outlining his VALUE. He attached only an "Executive Summary CV"—one page that was easy to read in seven seconds. He followed procedure and worked with the two executive assistants (Christine and Nicole) to set up calls with each of their bosses.

Christine and Nicole put Lance's "letter" on their boss's desk. Lance's message resonated with Christine's boss John, who instructed his assistant Bill to "let me meet the guy for no more than 10 minutes." John said, "This may be an opportunity to solve our problem." In a face-to-face interview later that day John quickly decoded the value offered and *created* a *new job* in his department. John told his boss that he had a great idea and, in fact, he found the perfect resource. Two days later Christine called HR and instructed them to meet with Lance on the term sheet. One week later Lance was hired. In the interim he also got an offer from Nicole's boss.

I asked Lance why his last "last resort" was the most effective. His answer may surprise you. He said, "I forwarded the broadcast letter I wrote using the techniques laid out in your book. I called when I indicated I would. They requested my full resume. I forwarded one that summarized my background under functional headings as suggested in your book." At our first meeting John indicated several things:

- ✦ He received literally dozens of resumes and phone calls weekly. He responded to none—ever!
- ✦ He had all his calls screened to reduce the number of job seekers who got through.
- ✦ He did not like being hounded with unsolicited phone calls or inundated with resumes.
- ✦ He told me that the main reason he took my call when I followed up on the broadcast letter was that my "approach was unique." I did not throw a resume at him hoping he liked what he saw. Nor did I call repeatedly.

Lance realized that he indeed *created* a new job that did not exist before by triggering the value requirement in John's company. "I spoke to his needs and goals, indicating that there may be a mutual advantage to us speaking for a few minutes," said Lance. In the end, Lance was *the ONLY applicant*!

✦ ✦ ✦

This new reality is likely how the job-search world will remain for the next five years in technology. Target marketing your skills will remain the *only* way to find meaningful work quickly.

Lance's story is backed up by a May 2003 survey by Perry-Martel International Inc. (my executive search firm) and the Canadian Advanced Technology Alliance, which is the largest association of technology companies in Canada. Sensing that hiring demands in the labor market had changed, we set out to interview a cross section of technology firms across North America. In total 19,000 technology leaders from across North America were asked to vote on the most effective way to find a job today based on *Parachute*'s original 10 ways. (The results of the 7,182 participants are shown in the following table, and a complete discussion of the survey appears in Appendix M.)

The 2003 Survey of High-Tech Employers found that if high-tech workers want to get a job, they have to dramatically change their approach to the search process.

The environment following the high-tech meltdown is one where an oversupply of workers is being accompanied by a shortage of highly skilled workers. This is causing the dynamics of

THE WAY to Find a Job (Post-2001)

		Pre-2001	Post-2001
The 9 Worst Ways To Find A Job	Randomly mail out resumes to employers: Most e-mail packages are set up to capture SPAM at the source, and your resume will be deleted before it's ever read.	7%	<1%
	Answer ads in professional or trade journals appropriate to your field: Employers will not advertise here, unless they want to build their database, because the ad will last a long time and the responses will come in for months.	7%	<1%
	Answer ads in newspapers in other parts of the state or country: You'll be in the pile of last resumes to ever be read because you're from "away." Forget about launching a geographic discrimination suit. It won't fly!	10%	<1%
	Answer ads in local papers: Cattle-calls and "resume drives" are popular ways for employers to build their databases. You have about a 1 in 3,000 chance of getting hired.	5–24%	<1%
	Use headhunters: Estimates vary but antidotal evidence concludes that 90 percent of all headhunters were out of business within two months of the recession starting in January 2001.	5–24%	<1%
	Personal Referrals: The telecom industry alone laid off 317,777 people in 2001. According to the September issue of *Business 2.0*, in 2002, one in five laid-off people in America worked in Telecom. It's tough to find a job through your network of colleagues when you're all looking at the same time.	33%	<1%
	Knock on the doors: "Walk-ins" work well for positions paying less than $30K. Walk-ins will get referred to the HR department. When a downturn comes hiring is frozen—period. HR's job is to hold the line and say no. The only people who can break "a hiring freeze" are "C"-level executives or managers with a particular problem.	47%	<1%
	Yellow Pages: Cold calling potential employers from the Yellow Pages is great when times are booming. Now it's a waste of time unless you can cold call a senior executive and explain exactly how he can make money, save money, or increase efficiency right then and there over the phone.	69%	<1%
	Job Hunter's Club: Not a chance...unless the group is truly made of selfless people who will actually "share" leads and tips openly. Otherwise you're likely to all be competing for the same jobs and the "survivor" instinct will kick in. It's actually counterproductive.	84%	<1%
THE WAY	Target Marketing: Find a company with a *problem you can solve*. Figure out how *your skill set can help solve its problems*. Articulate how you can add value (refer to the New Value Table in Chapter 1). Find out who has the authority to hire you. *Initiate a dialogue or network your way in.*	86%	97%

job hunting to change dramatically from the situation a few years ago. The old ways of job hunting not only don't work, they have been supplanted by a new approach.

Among the Survey Findings

"The ways of the world in 2003 are very different from the world of 2001," noted CATA president John Reid, co-sponsor of the survey. "The environment is sending contradictory signals: While many high-tech workers are seeking jobs today, employers are facing a shortage of highly skilled workers."

Sensing a shift in hiring practices, PMI and CATA set out to survey North American employers on their actual hiring practices and develop a guide to tell technology workers how to best find new employment in the post-dot-com high-tech economy. Employers were represented in two categories: executives and non-executives, with major differences in attitudes between the two groups.

There were no differences between industries or countries on the best way to find a high-tech job. The differences are between executives and nonexecutives.

Of executives, 97 percent feel that the best way to find a high-tech job today is through target marketing of firms followed by a direct approach to the executive. For executives, the second best method is through recruitment companies. "Networking" rated less than a 1-percent response.

Among nonexecutives, however, 73 percent felt that networking was the best way to find a job. This feeling was strongest among human resources professionals (83 percent) and lowest among the engineering professionals (57 percent favored going to recruiting companies).

Of executives, 84 percent said they were always in strategic hiring mode, versus 7 percent of nonexecutives.

Executives (81 percent) concentrate almost entirely on intangible skills instead of tangible skills. They look for core qualities that add value to the profit-line of the organization. Nonexecutives (77 percent), however, are still looking to fill in boxes on a standard recruiting form.

The biggest surprise in the survey was that networking rated so low among executives. Executives prefer a direct approach and are less likely to suggest using headhunters, whereas non-executives still believe networking is the best way to find a job. In the kind of hiring freeze situation that often accompanies a slow-down, it's easy to understand why nonexecutive staff may be loath to bring forward anyone they interview, no matter how strong the person's skill set may be. This means that the networking favored by nonexecutives will keep job seekers very busy but is not likely to result in a job offer.

The catch-22 in all of this is that, although executives are looking for people with the kind of qualities that can advance their business, their hiring staff has a very different set of standards for the candidates they think the company needs.

To resolve the imbalance between job seekers and executive searchers, both sides need to approach the problem differently.

As a job seeker you need to become adept at direct self-promotion to executives and to showcase your answer to this question: "How can I increase shareholder value?"

Employers need to instill in their search teams the values they themselves cherish in their continuing search for star talent—that is, the realization that value is not salary; worth does not flow from a job title. Knowing how to evaluate the worth of someone's contribution is the important element. But as a job-seeker you must change first because corporations are generally last to change.

(Results of the entire survey are in Appendix M. The entire piece deserves your attention.)

The most important thing for you to remember as a job-seeker is that employers are *always* looking for new talent. As is that of a football or hockey coach, a manager's job is to know the bench strength of his team and to always be on the lookout for new talent that covers the weak areas and augments the strengths. During times of economic uncertainty employers need to find the star performers who are critical to success. Today's economy is an environment for lean companies, driven by a relative handful of highest-quality employees. Tomorrow's economy will be even more demanding. When China ramps up as a global economy the profit margins in many companies will be squeezed to the max so every employee must be an "A" player.

Obviously, as a job-seeker you need to speak to employers in terms they understand. One of the major findings from the survey is what we refer to as the New Value Table (see page 27). This table speaks directly to an employer's need for innovation and wealth creation.

The New New Economy

For the last decade, if you could spell "computer" you were pretty much assured a high-tech job. The demand for high-tech workers was expanding exponentially. Resumes were a formality and often unnecessary. Interviews were one-sided employer sales pitches, and references were rarely checked. That all changed in January 2001.

The high-tech meltdown is driving organizations to be aggressive to obtain the best talent: the top people whose skills mean the difference between victorious and vanished. In this leadership-hungry environment, both recruiting and looking for a job have changed. For employers, it's the difference between standing on the street corner handing out leaflets and looking in all the company windows, finding those you need, and then breaking the glass to reach in and take them.

Employees who have talent are still very much in demand. Any person who can design a top product, manage complex projects, perform marketing miracles, sell new customers, or execute leadership is an authority who can take his or her pick of top opportunities. For job seekers, heavy preparation and due diligence are necessary to find which companies have the "team" and "insight" to make it, because you'll only want to make a move if the career attraction is compelling.

If you want to continue to advance in your career, you need to learn how to be your own headhunter, because the market is in turmoil and the calls from recruiters are far less frequent. You need to become adept at self-promotion. Learn how to search the world, cold-call prospects, get their attention, raise your proposition above the background noise, keep at it tenaciously for however long it takes—be it weeks or months—and be intelligent enough to present your skill set in creative new lights until the persuasion works.

Listen and Learn Now

Your skills and personal qualities define your career opportunities. There are no restrictions on what you can do next—except those you impose on yourself. The tech industry is in its infancy. Knowing what's important to technology companies and how to reposition yourself are the keys to your future success.

Today, the ability to produce value is the most important criterion for hiring, and it should also be your most important job search criterion.

Value is not salary; worth does not flow from a job title. Knowing what's important to a company means looking beyond job titles and compensation tables, especially in the world of high-tech, where sudden changes and uncertainty are the norm. For employers, knowing how to evaluate the worth of someone's contribution is the important element. For you, being able to speak of your accomplishments and articulate your "star performance" capabilities will define the difference between an average and an extraordinary career.

There are two elements you need to consider:

◆ Understanding thoroughly what the value contribution is that an employer wants.

◆ Understanding how to position yourself and articulate your unique value to an employer.

More than ever in our history, huge value is being leveraged from smart ideas—and the winning technology and business models they create. High-tech workers are on the edge of change, and they see better than anyone else how their worth—your worth—measures up.

Especially for its senior positions, a high-tech company is rarely looking to fill in a box on a standard employee recruitment form. High-tech companies are looking for something more nebulous and more important. They are looking for a senior person who can deliver a *quality*, not a quantity. Instead of filling in a box, you are looking to explode out from an open-ended, initiative-driven space. Qualities are difficult to find, measure, or test in an ordinary recruitment drive. And you don't find those qualities by searching for specific salary levels; the qualities that make up the New Value Table are money-resistant. We will talk about the New Value Table in detail in Chapter 1.

◆ ◆ ◆

Based on the many long discussions I've had with thousands of employers, let me assure you that looking for a job in high-tech is a waste of time. Jobs don't have "a future." You have "a future." Jobs are temporary in the new economy. Careers are now made up of a succession of jobs over your lifetime, not a lifetime at just one. Your future will only be filled with meaningful work worthy of your time and effort and in line with your interests if you take charge of your life and build your career one job at a time with a specific end goal clearly in mind.

This book gives you all the basic tools you will need to do exactly that. Here in the pages of this book I have reverse engineered the same strategies and tactics professional recruiters use to find world-class talent for their clients. I strongly urge you to read the book, do the exercises, find world-class employers, and go create your life yourself.

This Is the Book Headhunters Did Not Want Written!

In my 17 years as an executive search professional, headhunter, recruiter, and placement counselor, I have negotiated more than $150 million in salaries. I know what to say and how to say it to get employers to make hiring decisions. I know all the tricks of the trade to turn finding a job into a logical and systematic process—I discovered almost all of them the hard way—and I am going to share them with you so that you don't need to go through the same agonizing trial-and-error experiments I did.

But, and there's always a "but," I don't want you to waste your time and hurt my reputation if you're one of the millions of people every year who dream of "hitting it big," and becoming a "high-roller" by winning the next lottery, but who never actually buy a ticket! This book is definitely not for you. You actually have to get in the game and work. There are no silver bullets! I hope that's not you I hear groaning, "Then why did I buy this book, Dave?" Why? Because there is a well-honed process I can teach you if you want to learn it and work it—but you actually have to do the work yourself.

I tell you this now because in the course of having this book reviewed by 47 colleagues and dozens of complete strangers, it received mixed reviews. *All* the headhunters, human resources managers, and career counselors—as well as *all* the individuals who were out of work and earnestly looking for new opportunities—raved. I kid you not. I was genuinely surprised. They like the simple straightforward advice.

On the other hand, the two businessmen who reviewed my book thought it was too "heavy." In their opinion, candidates today are looking for easy answers and simple fill-in-the-blanks solutions. As Tony Patterson (publisher [ex officio] of *Silicon Valley North Newspaper*) pointed out, "Oh it's a great book David, but people are lazy. Did you know that 95 percent of the people who attend seminars or read self-help books don't do anything with their newly acquired knowledge past the weekend? They come back all pumped-up and it just fizzles. It's money down-the-drain!" Peter Kemball another businessman whose opinion I deeply respect, suggested, "Candidates won't do the work, won't find a job, and will blame you for their lack of success."

Frankly, they both made valid points. I really could not risk the bad press this book might cause. On the other hand, I also could not ignore the hundreds of friends and the thousands of referrals who had begun calling my office in early February 2001. That wouldn't be right either. So I decided to reverse-engineer the executive search process designed to identify, locate, and attract star talent for my clients. In the last 15 years, clients called off the only searches I didn't successfully complete (two of 877) due to their business changing. So that's the process, you are getting yourself in to.

You have to decide if you want to advance your career on purpose. If so, then this book is for you. I'll show you how to "headhunt" your next job and how to work effectively with friends, colleagues, strangers, and headhunters so they can help you too. This book is full of purpose. It was written to be followed much like a recipe—one that's being a closely guarded family secret for generations.

When people ask me for career advice (which, by the way, is not really what I do—I am paid to acquire executive talent for companies: people for jobs, *not* jobs for people), I always ask them what they would like to do next in their careers. It's a pretty simple question really, but it's one most people cannot answer. Many enthusiastically reply, "Oh, I'm open to any new opportunities." Now, to an executive recruiter this literally translates to, "I don't really know, but I'm not enjoying this job." Not the type of first impression you want to give! I'll show you how to answer that affirmatively instead.

With predictable results most people leave their lives to chance, drifting from one "job" to another to another in search of a career. These same people may spend more time planning their weddings than working on their marriages. This book is not for them.

Are you feeling a little uncomfortable right about now? Do you have a pit in your stomach? Well, I guarantee you this book will help you find your perfect job faster than anything else you've ever tried and your friends will ask you how you did it. Read on.

How This Book Is Set Up

I assume you want to find the right job as quickly as possible. The book is set up so you don't need to read it from cover to cover. I really suggest you do because the basic building blocks of the process work in a specific order, but sometimes you just need a specific piece of information right away because you are blocked somewhere in your search. To make things easier to find I've organized the book by the following stand-alone sections.

Getting Started

If you're just starting your job search after years of steady employment, start here. Take the time to analyze what you've done and what you want to do next. The exercises are designed to quickly determine your most marketable skills. Investing a few hours here will help you build a solid foundation for your resume, cover letter, and search criteria.

Your Resume

If you're not getting the results you want from the resume you've written, this section covers the how-to for dynamic resumes that reflect your accomplishments and make employers want to call you.

Correspondence With Samples

If you already have a great resume but can't articulate in your cover letters why you should get an interview ahead of someone else in a cover letter, this is the place to go.

How to Find a Job

If you're tired of chasing after the same jobs as everyone else you need to discover the "hidden job market." This section shows you how to use search engines and other *free* resources to do that quickly, putting you ahead of your competition.

The Active Interview: A Two-Way Situation

Do you have an interview tomorrow for a dream job you just found? Start here and find out how to set yourself apart from everyone else *and* determine if this is the kind of company that has what it takes to be successful, so you don't have to look for a job again!

Finding and Using a Recruiter

Recruiters can be your career lifeline but they aren't going to waste their time with people who are unfocused, unsure, or ill-prepared. Read this section if you want to understand and successfully work with the world's "job search commandos"!

Resiliency: The Winner's Edge

Straight talk about you and your need to really commit to the process and your job search.

Want More? Here It Is!

There are 14 appendices to help you speed the process. I've included most everything you'll need from a form on which to record your career history to sample resumes and a search engine "cheat sheet." Perhaps the most important though is the Job Search Organizer. It is the same type of organizer that recruiters use to manage and track their placement activities. It is a good idea to get into a regular routine, and the organizer does that nicely.

Still More?

I have set up a free Website to help. (Yes, *free*.) At *www.workinsight.com* you'll find, on average, 50,000 jobs that have been aggregated from technology newsgroups across North America. Please surf and apply to your hearts content.

Chapter 1

Getting Started

What's your life's mission? What are you really passionate about? Who do you want to work with? What type of work so absolutely fascinates you that you would do it for free? If you can answer those questions, you can start to build a career you will excel in—the best of both worlds!

If you don't already know the answer to those questions then start right here and carefully read the next few pages. You will quickly see that you need a foundation for your job search. No matter what your age or experience, if you have not identified your ideal job, then all the remaining steps (researching your career options, preparing your resume, conducting a job search, and interviewing) are very difficult to complete. But for unusual luck, it's impossible. In other words, searching without knowing what you are looking for becomes a dysfunctional hit-or-miss, try-before-you-buy experiment with your time and life.

On the other hand, once you understand what you would like to do and perhaps even the industry in which you would like to work, everything else falls into place rather naturally. Your research will be highly focused. Your resume will be relevant. You will network well. You will interview with confidence. You will get good job offers—more than one at a time! Maybe you will even provoke a bidding war!

On the other hand, if you loved the work you did in your last position and the industry in which you were working, then go straight to Exercise 1. If you did not, then consider doing the same or a similar job in a new industry. I often find that it's not the job but the industry that people don't connect with. Fortunately high-tech people tend to be the brightest on the planet, and whether you're in engineering, finance, or sales and marketing, many skills you have transfer easily from the high-tech industries to insurance, banking, automotive, healthcare, and so on.

Exercise 1 is designed to help you inventory your skills quickly and decide what you like to do. We will also rely heavily on it to help build your resume in Chapter 2 so you have to complete all of Exercise 1, which should take between one and four hours. (I took six, but I was double-checking my answers.)

The purpose of this first chapter is to help you identify your *ideal job* based on your skills/abilities, interests, and purpose. If you don't know what I am talking about when I say purpose, or want assistance in defining your purpose, the following two sites may be of real interest to you.

- *Fast Company*: *Deciding on Purpose*
  www.fastcompany.com/online/13/ldrplus.html
- Steven Covey's (*7 Principles of Effective People*) interactive tool
  www.franklincovey.com/missionbuilder/index.html

Exercise 1: Defining the Content of Your Resume

Have you ever heard the expression, "So what have you done for me lately"? I'll let you in on a little secret: Employers generally hire people for one or more of these three reasons:

1. To increase their profitability.
2. To decrease their costs.
3. To improve their efficiency.

Keep that foremost in your mind as you go through the book and these exercises specifically. You need to start thinking of how you can increase shareholder value. (More about this later.)

Beginning with your latest job and working backwards in time, describe all the full-time and part-time jobs you have ever had. Go ahead and mention the paper route when you were a kid, if you must, but it probably won't figure heavily in your resume, so don't get too carried away. If you can't remember all your responsibilities for a particular job, it doesn't matter. They couldn't have been that important.

List all your responsibilities for each job. When you're finished, list your accomplishments. I would like you to adhere to "Perry's 80/20 Rule"—that is, spend 80 percent of your time listing accomplishments and 20 percent of your time listing your responsibilities. For the purposes of getting a job, remember that employers care about what you did (accomplishments), not what you were responsible for. Employers are only interested in three things:

1. Can you make me money?
2. Can you save me money?
3. Can you increase my efficiency?

Being relevant to an employer's needs is key. It is what will get you noticed and hired. I want you to think about your accomplishments with the employers' value requirements in mind. You are going to focus all your efforts on addressing potential employers' needs. Your needs come into play after you get an offer. The New Value Table appears on the following page.

The New Value Table is an emerging concept, although it probably seems familiar to you. Like everything evolutionary, it has a certain "I know that" familiarity. It was derived

The New Value Table

Employer's Value Requirements	Your Quality That Counts
Create new intellectual wealth for my company; add to my intellectual assets.	A consuming desire to make something new, to cut a new path rather than take a road.
High-energy enthusiasm for the job, regardless of the hours worked.	Work is a game—an integral, vibrant part of his or her life.
Not only is money not the most important issue, it's beside the point.	Internal pride to leave a "legacy signature" on their work, rather than strive for a paycheck.
Enduring performance.	An ability to stay and finish the race, because not finishing is inconceivable emotionally.
"Think around corners" to solve problems creatively.	Have an inner voice saying, "There's always a way (to create a technology fix, make a deal, etc.)."
Bring up-to-date professionalism into every fray.	Contain a desire to grow professionally, to become the best person he or she can be; invest in themselves.
Ever-increasing contribution.	The key to inner pleasure is recognized as making an individual contribution.
Identify and develop values for your company.	Instinctive grasp and exploitation of today's real value: the intangible capital of brand image, staff talent, and customer relationships.

from the thousands of hours I spent talking with employers about how to grow their businesses through their people.

Here are some general questions that will help you focus on and articulate the qualities you possess that count for an employer. In your current/previous positions have you:

- Saved a faltering product or launched a new product? (Did you do something unique? Was it a product which management had forgotten or dismissed? Be careful not to paint yourself as too much of a rebel here.)
- Handled a special assignment outside of your official job? (Try to pick those that either added a great deal of value to the firm, or are relevant to the position for which you are applying, or show you understand how to advance your career yourself).
- Identified and solved problems that had been overlooked? (What's the story? Why where you successful?)
- Increased sales? (Be prepared to talk about who and when at the interview.)

- "Closed" an impossible client? (How come you got through? Did the client just have a bad day, or did you discover a technique you can teach others in the company?)
- Attracted new financing? (When, why, and how much?)
- Saved the company money? (Really? How much? Can you do the same for this company? Do you have a pattern of doing this at other companies?)
- Increased efficiency? (What was the end result?)
- Suggested an innovation? (Results?)
- Bootstrapped a new product line? (How did it do? Is it still alive today? Can you make parallels to the positions for whcih you will be interviewing?)
- Opened a key account? (Be careful with this one! I once had a colleague brag that he had opened more new accounts than anyone in the history of our company, but he hadn't closed *any* of them and when this certain employer called our boss—well, let's just say he laughed for days. Don't write it down if you can't back it up.)

When you write out your accomplishments, be detailed and very specific. You will be using this information to deliver bulleted messages in your resume, in your cover letters, and during the interview. This is the most important step in the process. Everything else you do will flow out of this exercise. Take your time. Don't be shy to ask colleagues for their input as well.

Being specific about your accomplishments means asking and answering questions such as:

- How much did you increase sales?
- During what time?
- By overcoming what hurdles?
- What were the "hard numbers" and percentages?

That's what will go into the resume. Write everything down: dates, supervisors' names, everything you can remember. We will worry about cutting down the volume of information Chapter2.

Keep these notes; you'll need the details handy when it comes time to interview. Most companies have adopted some form of "behavioral interview" format, which emphasizes drilling down on specific behaviors to distil individual actions. Recruiters tend to analyze accomplishments this way, so you will be pressed for details and, if you have to stop to think, your answers may be suspect. In fact, I once read somewhere that for every two seconds you spend thinking about an answer, you lose credibility with the interviewer. Frankly, I know I am suspicious when candidates can't tell their story smoothly. It could cost you a great job. Interviewers expect you to be fully prepared—whether that's realistic or not is irrelevant.

Exercise 1.1: Full-Time Work

Detail your career history now. You'll find a copy of the form to use in Appendix A. Take it and photocopy as many copies as you need.

- Employer: The legal name of the company and the city/location.

- Dates: Your start and stop dates (with years and months).
- Title: Your exact title and to whom you reported.
- Responsibilities: Listed in bullet form.
- Accomplishments: Listed in bullet form, using percentages and dollars.

Exercise 1.2: Volunteer Work

Volunteer work experience counts for a lot more than you might think. For one thing, employers like to see that you are a well-rounded individual and not just another cog in the wheel. Human resource managers use information such as this to assess your "fit." So it's important to touch on it, ever so briefly, in your resume.

Personally, what I learned at university pales in comparison to the skills I picked up as an officer in the Army Reserve. The "logistics training" I received alone would have cost my employer tens of thousands of dollars. Leading 120-plus people to the Arctic on an "outward-bound" expedition really honed my leadership skills. I think that experience more than any other paved the way for my first real job out of university. Managing a $4-million retail store for Consumers Distributing (Service Merchandise in the United States) was a piece of cake compared to canoe-sailing the Mackenzie River to Tuktoyuktuk and running from brown bears.

Many people gain management and leadership skills from volunteer activities through civic and religious organizations. Indeed, these types of opportunities provide a place to hone your non-technical "soft skills." You can develop skills through volunteering that you could never develop through working. In the United States, for example, very real technical skills can be picked up in the military reserves or National Guard. It's very comforting for most employers to know you've had "disciplined" training of some type when you're starting your career.

Not too long ago, McDonald's restaurants trained more people every year than any other employer in America. More importantly their training is universally recognized as solid. Several years ago I interviewed a director of customer service for a high-tech firm simply because she had been trained at Disneyland. She did not have the number of years of experience we were looking for, but we understood that her experience at Disneyland had given her a solid foundation. She has proven to be remarkable hire for my client.

List all your volunteer experience in detail with your accomplishments. (See Appendix B.)

- Employer: The legal name of the company and the city/location.
- Dates: Your start and stop dates (with years and months).
- Title: Your exact title and to whom you reported.
- Responsibilities: Listed in bullet form.
- Accomplishments: Listed in bullet form, using percentages and dollars.

Exercise 1.3: Education

List your highest levels of education in descending order. Exclude high school. Follow that with *all* your other training that is related to your *career*.

Post-Secondary: _____

Years attended: _____

Degree/Diploma/Certificate: _____

Other (that is, certifications, seminars, continuing education, etc.): _____

Exercise 1.4: Hobbies and Interests

People believe hobbies and interests reveal a lot about you and usually reflect your individuality. Personally, I don't fit the "stamp collector" profile. I bought and sold them to pay for most of my university education. More importantly, for our purposes, hobbies and interests showcase your interests outside work and are often the first things an inexperienced (90 percent) interviewer comments on during an interview to "break the ice."

List your hobbies and interests.

_____

_____

_____

_____

Exercise 1.5: Patents/Trademarks/Copyrights

In reverse chronological order, list all of your patents (indicate the patent number), trademarks, and copyrights on a separate piece of paper. (We'll get back to this later.)

Exercise 1.6: Papers and Publications

In reverse chronological order, list on a separate sheet of paper all of the papers and/or speeches you've delivered in the past 15 years, that are work-related. Do the same for publications. There will be a special resume page for these.

Exercise 2: Analyzing Your Skills and Personality Characteristics

Skills Index

Referring to the two skills lists that follow, review what you have written about your accomplishments. Highlight all your skills and personality characteristics. These are the main skills a high-tech employer will pay you for. How do you measure up? Which are your best? Highlight them now, because that is what we will play up in your resume and correspondence.

List of Functional Skills

Identify five to 10 functional skills that you can illustrate through your personal experience. You can also use them in writing your resume as *action words* or during a job interview to explain accomplishments. (See Appendix C.)

Administer	Budget	Evaluate	Persuade	Speak
Advise	Compile	Facilitate	Plan	Supervise
Analyze	Coordinate	Investigate	Recruit	Teach
Appraise	Counsel	Lead	Research	Write
Audit	Create	Manage	Sell	

You may want to write out examples now so you don't forget, because you'll need them later for interview preparation.

Now do the exact same exercise for the following personality characteristics.

Personality Characteristics

Ambitious	Daring	Humorous	Patient	Sharp-witted
Accurate	Deliberate	Imaginative	Perceptive	Sharp-minded
Active	Dependable	Independent	Persevering	Sincere
Adaptable	Determined	Individualistic	Pleasant	Sociable
Adventurous	Dignified	Industrious	Poised	Spontaneous
Aggressive	Discreet	Informal	Polite	Stable
Analytical	Dominant	Intellectual	Practical	Steady
Artistic	Dynamic	Intelligent	Precise	Strong
Assertive	Eager	Introspective	Productive	Strong-willed
Astute	Easygoing	Inventive	Progressive	Supportive
Bold	Efficient	Kind	Prudent	Tactful
Broad-minded	Emotional	Leisurely	Punctual	Teachable
Businesslike	Energetic	Light-hearted	Purposeful	Tenacious
Calm	Enthusiastic	Likeable	Quick	Thorough
Candid	Empathetic	Logical	Quiet	Thoughtful
Capable	Expressive	Loyal	Rational	Tolerant
Cautious	Fair-minded	Mature	Realistic	Tough
Charming	Firm	Methodical	Reflective	Trusting
Cheerful	Flexible	Meticulous	Relaxed	Trustworthy
Clear-headed	Forgiving	Mild	Reliable	Unaffected
Clever	Forceful	Moderate	Reserved	Unassuming
Competent	Formal	Modest	Resourceful	Understanding
Competitive	Frank	Motivated	Responsible	Unexcitable
Confident	Friendly	Obliging	Retiring	Uninhibited
Conscientious	Generous	Open-minded	Robust	Verbal
Conservative	Gentle	Opportunistic	Self-confident	Versatile
Considerate	Good-natured	Optimistic	Self-controlled	Warm
Consistent	Growth-oriented	Organized	Self-reliant	Wise
Cooperative	Healthy	Original	Sensible	Witty
Courageous	Helpful	Outgoing	Sensitive	
Curious	Honest	Painstaking	Serious	

Exercise 3. Listing Your Technical Competencies

If you are looking for a technical job, you need to produce a special section in your resume just to display your technical skills. Here's an example addendum:

Technical Summary

Systems development methodologies/tools

Information Engineering
Workbench (IEW)
Productivity Plus (DMR)
Method/1 (Andersen Consulting)
Manage/1 (Andersen Consulting)
Change/1 (Andersen Consulting)
Information Engineering Techniques
Integrated Systems Methodology (PWC)

Application/Database

Microsoft Products (Mail, Project, Word)
Lotus Products (Notes, 123, Ami Pro)
Ingres
Oracle
Excelerator
WordPerfect
Graphics Products (Harvard, etc.)
ZIM
FoxPro/Clipper/Dbase
Natural/Adabas
Data Manager

Internet

Browsers
Electronic Mailers
News Readers
Search Engines

Hardware

HP 9000/T500 UNIX RISC
IBM/Macintosh PC
IBM 43XX, 30XX, 1130
IBM System 360/370
DEC VAX
NCR 8250/9020
Honeywell/Singer

System software

Windows 95/3.11/2000
Unix/Xenix
MSDOS/PCDOS
Jes 3/MVS
OS/VS
IMOS
TSO/ISPF
Complete
Panvalet
Job Control Language (JCL)

Languages

C
COBOL
Natural
Fortran
RPG
Assembler
TPL/PCL
Basic

Technology Skills

Referring to the following list (which is by no means definitive) of "hardware" and "software" skills, highlight all that apply. At the end of Exercise 3, list all the other software and hardware skills you possess.

Software Skills

Ada	Assembler	C/C++
CGI	COBOL	Crystal Reports
Delphi	FileMaker Pro	Fortran
HTML	Java/J++	JavaScript/JScript
JCL	LISP	Oracle
Pascal	Perl	Prolog
SAP	SQL/ODBC/JDBC	Visual Basic
MFC/Visual C++	XML	Motif
VRML		

Special Software Skills

Artificial Intelligence	Business Process Reengineering
Concurrent Programming	Data Warehousing
E-Commerce	EDI
Graphics	GUI
Low Level	Networking
Object Oriented Design	Operating System Development
Real-Time/Embedded	Strategic Planning
Systems Programming	Testing
WWW Related Skills	

Protocols

Apple Talk ATM	COM/DCOM	CORBA	CMIP
DDK	EDI	HTTP	LU6
NetBEUI	OSI	SDK	SMP
SNA	SNMP	SS7	TCP/IP/UDP
X.25	X.400	X.500	XIP

Database Skills

Adabase/Natural	Access	DB2
Dbase III	Dbase IV	Filemaker Pro
Fox Pro	InterBase	Informix
Ingres	Lotus Notes	Oracle
Paridox	Progress	Power Builder
SQL	Sybase	Starbase
ZIM		

Operating Systems

AIX	DOS	Digital Unix	HPUX
Linux	Macintosh	MFC	MOTIF
MVS	Netware	OS2	SCO
SGI	Sun/Solaris	Tandem NSK	Unix
VAX/VMS	Windows 3x	Windows 9x	Windows NT
X11			

CASE Tools

Bachman	Composer	Endevor
Excelerator	IEF	IEW
Navigator	Rational	

Hardware Skills

Analog	ASIC	CMOS	DSP
Digital	Networking	PCB	RF
RSP	VLSI		

CAD Skills

AutoCAD	Cadence	Exemplar
FPGA	LSI Logic	Logic Work
Maple	Mentor Graphics	MATLAB
Model Tec	Neocad	View Logic
XACT Tools	Xilinx	

Non-Technical Skills

Team Leader	Project Manager	Project Management
Project Planning		

General Skills

Architecture	Application Tech Support	Business Modeling
Change Management	Course Ware	Database Development
Joint Application Design	Programming Development	Process Reengineering
Quality Assurance	Requirements Analysis	Technical Writing
Bottom of Form		

Now, on a separate sheet of paper, list all the other skills you think are relevant here. When I say relevant, I want you to put yourself in an employer's shoes and ask yourself, "Would I pay for someone to bring that skill with him or her to work for my company?" For example, formatting floppy disks is a skill but not one an employer is likely to pay for; neither is defragging a hard drive or making coffee.

Exercise 4: Defining Your Preferred Skill Summary

Let's start to pull this together. Right now you've probably got 10 to 30 pages of information, and at this point you need to decide which skills you prefer to work with. Referring to Exercise 1 and Exercise 2 *but not* Exercise 3, jot down the skills *you* value most.

Next, circle those skills an employer is most likely to see value in (in other words, skills the company will pay money for). Put yourself in the employer's shoes: when money is tight, what skills would you be willing to buy (for example, java vs. COBOL; sales vs. shipping)?

Lastly, of those skills you highlighted on pages 33 and 34, which are circled as well? The highlighted and circled skills are the skills that you like to use the most *and* are also the skills employers are likely to pay you for. Remember: Employers will only pay you for what you *can* do, not what you studied to do. Ideally what you *can* do is also what you *like* to do. With a little effort on your part, your next job can reflect that.

On a separate sheet of paper, list the preferred skills and abilities that you want to use in your next job.

Exercise 5: Choosing Where You Want to Work

Did you know you spend as much time working as you do sleeping? Did you know that many professionals in the mental health industry think that job fulfillment is the key to your mental and financial well-being? Personally, a challenging job with fun people in an environment I like is mandatory. Given the explosion of opportunities caused by technology in the past 10 years, there's no excuse for not enjoying your work.

In order to find a job that really matters to you, first you need to know what motivates you and what deep subliminal "needs" you expect to have met by work.

Here's a quick way to do that. Underline every factor that is an important motivator for you and then go back and circle your top five.

A. **Working with:**
   - Data.
   - People.
   - Things (such as computers).

B. **Working for a:**
   - Large company.
   - Medium-size company.
   - Small company.
   - Start-up company.

C. **At a job that offers:**
   - Upward mobility.
   - Freelance work.
   - Decision-making authority.

- Fellowship on the job.
- Flexible working hours.
- Indoor/outdoor work.
- Opportunity for advancement.
- Opportunity to work independently or as part of a team.
- Recognition.
- _____ (Anything else a job should offer to hold your interest.)

D. Compensation:
- Salary (minimum): $_____.
- Commission: _____.
- Other (define): _____.

E. My basic needs and values are to: (highlight your five most important)
- Acquire money/material things/status.
- Be a leader.
- Be an entrepreneur.
- Be creative.
- Be financially secure.
- Be independent.
- Combat/prevail over adversaries/evil.
- Enjoy challenge, risk-taking, adventure.
- Excel.
- Exploit potential (situations, markets, people).
- Gain recognition.
- Grow as a person (pursue knowledge, understanding).
- Have many friends.
- Have time for leisure activities.
- Help other people/organizations/causes.
- Influence policy/people/attitudes.
- Organize and operate—business/team.
- Overcome/persevere.
- Provide a valuable service to society.
- Spend time with family.

You may be surprised to find that "psychic cash" is more important to you than real money itself. Everyone has a minimum they'll roll out of bed for. Most people even know what they'd like to make. Many people, however, are not aware of the importance of being appreciated for their contribution, but deep down people want to believe that what they do matters and is worthy of their time. If you examine what you highlighted in sections C and E, you'll begin to understand the filters you'll need to apply to *all* the opportunities you look at to find *one* you'll enjoy doing for the next 10 years.

Exercise 6: Identifying Ground Zero

Here's where this all comes together for you. Write a 40-word paragraph on the type of company you want to work for and the conditions you want to work in.

This is a practical exercise, because you will be using this paragraph as a guide to choose which companies and opportunities to pursue.

Exercise 7: Defining the Industries of Interest to You

Technology transcends many industries and work environments. For example, I sit on the board of directors of the Software Human Resource Council and I'm vice-chair of the Canadian Technology Human Resource Board. In both cases it's often very difficult to describe our "stake holders" because software, as is technology, is everywhere.

Technology is so pervasive that I can't think of *any* industry that doesn't use it. Looking around your own city, you'll be surprised how many work environments and industries from which you have to choose.

I have listed 38 industries here. Each offers multiple opportunities. The electronics environment, for example, might include jobs in sales, manufacturing, marketing, accounting, and R&D. The list is by no means complete. For more ask your local librarian for a copy of the SIC codes (Standard Industry Classifications).

Go through this list twice. First, draw a line through each industry that is not of interest. Second, circle the five you *are* most interested in. List all the job titles you can think of that use your skills. You should be able to list three to seven jobs in each industry.

Accounting	Defense	Investments	Software
Advertising	Education	Law	Telecom
Aeronautics	Electronics	Medicine	Television
Agriculture	Entertainment	Military	Textiles
Architecture	Grocery	Music	Toys
Automotive	Healthcare	Paper	Transportation
Banking	Hospitality	Photography	Wireless
Communications	Industrial design	Publishing	World Wide Web
Computers	Insurance	Real estate	
Construction	Interior design	Retail	

Arrange all your piles of written documents neatly, because we'll be using them in Chapter 2 to build your resume.

Your Resume

There have been thousands of books written about the correct way to compose a resume. I've personally read more than a hundred. Although each author has always made valid points based on his or her experiences, I have yet to read one written by a professional recruiter or headhunter. And who's better qualified to tell you about what gets read and what doesn't than someone who makes a living reading and assessing them every day?

My advice comes from working on that side of the job search process where I get paid to make professional hires every day of my life. Mine is a unique "value-driven" perspective where I don't eat unless I am right 99 times out of 100.

In my opinion, people forget that looking for a job is a sales and marketing exercise and your resume is simply an ad designed to get you an interview, where you "sell" yourself. All successful ads highlight a product's features and emphasize the benefits the buyer receives. Most resumes, however, are written like feature-heavy glossy brochures, historical novels, or, even worse, owner's manuals—leaving the readers to distil the benefits themselves. That's so wrong.

Like an immature copywriter, most writers and do-it-yourself job-seekers mistakenly think that the more features they pack into the ad, the better the chance someone will see something he or she likes. That's wrong, but it certainly explains why, despite the tens of thousands of brochures written every year, companies still need to employ a direct sales force to sell their products.

The technology boom of the last 10 to 12 years has produced an entire generation of job seekers just like you who know little about the fine art of resume preparation. Let's face it: For the past 10 years, if you had a degree and a pulse, some recruiter somewhere would do all the heavy lifting for you: Produce your resume. Market your skills. Negotiate and close your offer. Send you flowers on your first day or maybe even a gold pen. It was a war for talent.

You were courted heavily from all sides during the "tech boom." Headhunters used your resume and their keen psychological insight to strategically heighten an employer's interest, secure an interview, and get you offers in a market in which demand wildly outpaced supply. It was a battle for brainpower, a war for talent.

The war for talent around the world has evolved in recent years but still continues unabated. It has morphed, however. In its first incarnation, it was a quantitative battle best described as the "war for any talent." Today, it's a qualitative conflict, characterized as the "war for the best talent," by Peter Weddle in his book *Generalship* (Weddle's, 2004).

However, the war for the best talent has different objectives, different players, and different rules. What's the biggest change for you as a job seeker? As much as 50 percent of America's headhunters are extinct. Your career is in your own hands. Today, *you must* write a resume that positions you as a can't-do-without resource, because it's likely that you lack the verbal acuity and moxie to cold-call an employer and present yourself in a compelling enough manner to force an interview.

In this section we will be pulling together all the pieces from Chapter 1 to build a high-impact resume that will make employers call you.

Resume Basics

Your resume must follow only a few real rules without fail. Your resume must be typed—not handwritten. You must use new paper—not the blank side of some other document. (Trust me on this. You won't appear thrifty to an employer; you'll be singled out as incredibly naïve, at best.) I strongly suggest you stick with a bright white 24-pound paper if you can afford the few extra dollars. Do not put your resume in any type of folder to get more attention. If the resume is more than one page then attach one of those baby blue legal corners to the upper left hand corner. (Legal corners are cheap and are available at any stationary store.)

In independent tests resumes with blue legal corners where selected first out of piles of resumes 57 percent of the time. By comparison *all* the resumes in folders were set aside to have someone take them out of the folder later. You want to have your resume found in the recruiter's first pass—not the last, as it may be too late.

How You Will Use Your Resume

Think of your resume a strategic marketing document that will be used in a variety of ways:

- ◆ To advertise your availability.
- ◆ To answer newspaper ads.
- ◆ To cut and paste from to fill out an online application form.
- ◆ To rehearse before interviews.

- To draw interviewers' attention to a particular accomplishment during interviews.

- To tailor a thank you letter after interviews.

- As an aid for your references so they'll remember what you did, especially if you were one of many people on a large team.

- As a "cheat sheet" during telephone interviews because, after all, the interviewer has a copy.

Types of Resumes

There are two main resume formats to choose from: chronological and functional. Each has its place. You would logically choose whichever is better for you. I prefer a hybrid myself, which is value-based and achievement-driven. When I have a friend who's competing with a large number of job-seekers I often suggest an eXtreme resume makeover (more on this later, including examples).

Whichever format you choose, you will create three versions of it:

1. Plain (for direct approaches to employers, networking, headhunters, and so on).
2. Containing a "Job Objective" box (for responding to newspaper ads and Website postings).
3. Able to be scanned (for resume banks, job boards, and employers' online forms).

Chronological

This is the most commonly used format and the one many employers like because it is easy to follow and does not require any thinking skills on behalf of the reader.

Chronological resumes are the most effective if you have a consistent work history that readily demonstrates a logical career progression, helping the reader easily connect the dots.

These resumes are especially useful if you intend to stay in your current industry, because it shows the reader exactly what you've done and where you fit by detailing your most recent experience first and then working backwards through your career history.

They are not ideal for demonstrating accomplishments because people rarely read past your last two jobs before they decide to screen you in or out. You always want to be screened in. So if you use this format make sure you pull all your accomplishments to the front page. The higher up on the page the better.

Chronological Example

Mr. A
123 Daly Crescent
New York, New York
(212) 555-6995
ma@abcd.com

OBJECTIVE
Senior Systems Engineer in a fast moving small to mid-sized company. Preferably including international exposure and/or travel.

KEY ACCOMPLISHMENTS
- Restructured and improved the IT department infrastructure of a midsize, third world, African organization. Upon which we built a large Management Information System (MIS).
- Started up and ran an Internet consulting firm.
- Founder of Verisim, Inc., a massively multi-player, continuous time, on-line gaming company.

EDUCATION
University of Waterloo, Waterloo, ON
B.S., Computer Engineering, 1994
Courses included: Entrepreneurship, Software Development, Control Systems and Digital Design.

EMPLOYMENT HISTORY
MASAF, Malawi, Africa
MIS/IT Specialist, 1999–2001
Under adverse conditions, I revamped and improved their IT infrastructure and oversaw the development and implementation of a Management Information System. IT infrastructure improvements included a reliable computer network, full email and Web access, centralized backups, and a responsive IT department.

Functional

A functional resume takes your skills and accomplishments and then groups them into functional areas (for example, sales, marketing, technical, finance, management).

Functional resumes work best when you are planning a career change—hence the often-heard and much over-used term transferable skills. With a functional resume you demonstrate how your skills and accomplishments can be mapped to another industry. I successfully used a functional resume to take myself out of the retail industry and into banking.

Functional resumes also work well for people who have changed employers frequently (more than two moves in three years) or are returning to the workforce. Functional resumes downplay your movement (hence the stability concern that human resources people always have is mitigated somewhat) and emphasize your accomplishments.

Functional resumes are also a traditional way for people with "start-up" experience to readily demonstrate their value AND NOT be penalized for being entrepreneurs.

Functional resumes work particularly well for sales and marketing people.

Functional Example

Mr. A
123 Daly Crescent
New York, NY
(212) 555-6995
ma@abcd.com

SUMMARY

Over ten years' progressive experience in human resources, marketing & consulting, business development, product development, project management and sales to local and national private and governmental organizations. Leadership ability developed in the successful formation, focus and guidance of new ventures.

MAJOR ACCOMPLISHMENTS

Consulting
• started and successfully developed a human resources consulting firm (Sept.1988) which specializes in executive search and strategic human resources planning. Firm has received national attention in our specific market niche and increased revenues by at least 40% each year since 1988.

Business Development
• planned and implemented the market entry strategy for a new consulting division, including; primary market research, target market definition, pricing and service literature.

• led proactive business development campaign in Toronto and Montreal which resulted in the acquisition of four major blue chip accounts from direct competitors, subsequent revenues exceeded 250k and accounted for 37% of the firm's annual revenue.

• recommended and initiated the formation of a new profit center resulting in complete domination of the local Ottawa market and a strong national presence.

Value-based

I am constantly asked by friends (and complete strangers at parties) for my opinion on their resumes. You see, I'm "in the business" and therefore a perceived expert. With few exceptions, the typical resume rarely reflects a job seeker.

At parties, I usually comply because I'm polite. My wife, however, often gets left on the dance floor or, worse, gets scooped away by someone more interesting than I am. This was a persistent problem for years, so being a nice guy and not wishing to spend my life sidelined at parties, I designed my own example that people could mimic. Not surprisingly, it's achievement-heavy and based on the New Value Table (page 27). Accomplishing this was rather simple. I combined my chronological career history with a functional accomplishments section. It has been used thousands of times, modifies only a few rules, and delivers to the reader what he or she needs to know in the first six seconds to have the entire resume read. It's also easy on the eyes. (See Appendix D for a complete value-based resume sample.)

A value-based resume is powerful because it addresses the one question on every single employer's mind: What can this candidate do for me? This concise no-nonsense approach transmits your bias toward action and results. It says, "Here's what I've accomplished so far. What can I do for you?" Employers immediately grasp the depth and breadth of your experience because, as they do, you have passion and a bias toward action.

Obviously, you'll have to decide which format is best for you. Whichever style you choose, here is the content that must go into your resume.

The Content of Your Resume

Your Coordinates

Your reader needs to be able to reach you—promptly. Put your name, address, phone number(s), and e-mail address at the top of the page. Logical? You have no idea how many people forget this or assume their cover letters will stay stapled to their resumes. For most recruiters, this simple intelligence test separates "real candidates" from the "might-have-beens" rather quickly.

It's crucial that, if you are going to give out your office number, you can talk when people call. Cubicle life can be hazardous to your health if your boss is next door and you are whispering with a headhunter.

If you give out a home phone number that other family members use, please be sure your children are polite when they answer the telephone and ensure that they know how to take a message. First impressions last, and rude children make a very bad impression. You would be amazed at the things I've heard from bratty kids.

Job Objective

There is a debate raging among recruiters and HR folk about job objectives. Supporters and detractors are from both camps. In my opinion most "job objective" statements either pigeonhole you into a fixed category or are so vague that they're meaningless.

Job objectives are *great* when you are replying to a specific newspaper or Web ad. Simply put the ad's title (Sales Manager, CTO, etc.) below your name on your resume's first page and voilà!—a perfectly tailored job objective. There is no need to elaborate.

From my vantage point as a recruiter, I can see that the major problem with a job objective section is that people try to make it fit a very broad category of job opportunity. They adopt a "one-size-fits-all" mentality, I believe, because they're lazy. The end result is a statement that is so vague that it leaves the reader wondering if you have even the foggiest idea of what you want to do. And believe me, if you don't know what you want to do, no one is going to take the time to tell you. Employers will move on to the next resume and the next until they find someone who speaks directly to their needs.

If you want to include a job objective elsewhere than where I have indicated, get very specific. Here's an example.

Original:

> *Objective: A sales management position where my interpersonal, communication, and technical skills will be utilized toward continued personal growth and career advancement.*

Revised:

> *Objective: A Sales Management position where my 15 years experience selling to Telcos will increase shareholder value.*

The first objective I'm sure sounds great to the person who wrote it, but as an employer I don't care about your "personal career growth and career advancement." I have a problem to be solved, and either you can help me solve it or I'll hire someone else. The second candidate "will increase shareholder value"—now we're both talking the same language. John F. Kennedy put it best many years ago during his 1961 inaugural address, "And so, my fellow Americans: Ask not what your country can do for you—ask what you can do for your country."

So take a fresh look at your objective *and* your whole resume. Are you focusing on you and your needs? If so, your job search will take longer than it should. *Focus on the employer and how you can help solve its problems.*

Experience

This section can also be called "Work Experience," "Work History," or "Employment." Personally, I prefer "Career History," because I don't work; I create value. Semantics? Decide for yourself.

This section should include the company name, your job title, dates of employment, and major accomplishments. List your jobs in reverse chronological order.

One recruiter friend recently reminded me that she prefers a line about who the company is when the company is not well known. For example, everyone knows who Nortel and IBM are, but XGY Corp.? I don't think so. Help the reader position your company and your experience relative to your industry. It's in your best interest.

Being the president of ABC might eliminate you from a sales manager job at XYZ—unless you've indicated it was an "SME focused on software sales in Dallas." Now you've gone up a "notch" to the reader, because you've also demonstrated your entrepreneurial drive. Without that line, your resume would likely have been tossed aside because you would have been perceived as having too much experience.

Please, only take credit for what you've done. If you were part of a team, indicate so. You'll get points for being a team player without having to say, "I'm a team-player" *and* you get bonus points for being honest. Besides, you don't want to risk getting caught in a lie when employers check your references, because they will withdraw their job offer.

Finally, make sure you use the list of action words from Appendix C.

Education

Organize this section by listing all your postsecondary degrees in reverse chronological order. Keep it to degree, college or university name, city, and year graduated. *Never, ever list your courses.* Resume real estate is at a premium, and you can't afford to waste it. You will get hired based on what you've done and what you want to do, not what courses you've taken. There are few exceptions to this rule.

If you have more than five years experience, then high school is left off when that's all you have. Additional training or special courses that are relevant to the job that you are seeking are listed next. Include high school if that's all you have.

Don't put, "One course shy of an MBA." Whenever I see this, my first reaction is "stupid." My second is "lazy." It's all downhill from there. There is nothing you can possibly gain from listing this on your resume.

Recently, a client and I were deciding on two very strong candidates for a senior finance role. Both were accomplished and professional but the client could not get past the fact that that candidate B had indicated on his resume that he was one course shy of his MBA and had been for more than 10 years. It screamed project incompletion to him. We chose the other candidate, and this fellow lost out on a $150K opportunity.

You might ask why I had not warned the candidate ahead of time when I had, indeed, read the resume and done the initial interview. Well there are two reasons: First, it's not my job to doctor or alter a candidate's resume (and in my opinion that practice should be illegal). Secondly, the candidate felt it was important that we know he had not completed the coursework or he would not have put it in. Why? We don't know. Perhaps he was trying to warn us about a weakness. It doesn't matter. He didn't get the job because of that slip.

Also, do not list any part-time courses you took to become a commodities trader or home inspector unless they add value to the job you are after. You'll simply appear wishy-washy at best or a "quick-buck artist" at worst.

Keep the education section focused on and applicable to the type of job you are seeking.

Affiliations and Associations

If you have an industry association or other "professional group" of which you are an *active* member, by all means, list it. Religious groups are often frowned upon, so you need to think that through carefully. Leave any political and/or sexual cause-based groups off your resume. Nobody cares, and it's not relevant to your job.

Patents/Trademarks/Copyrights

List all that you have. It's quite acceptable to have a separate addendum for them. Organize them in reverse chronological order. For patents be sure to include the patent number and the year it was filed.

References

There *should not* be a section called "References." Do not list your references. I could write pages on why you must not do this. Don't do it.

Also, "references available upon request" is redundant. Everyone knows references are available upon request. What's more, you run the risk of having recruiters and executives brand you as "light" if you use that phrase—defeating the *whole purpose* of this book— because employers already know you'll provide references when asked. Saying so signals to many interviewers that you're an amateur job seeker.

The Unwritten Rules of the Game

There are certain unwritten rules that if adhered to will increase your chances of getting the attention you deserve.

Give Your Readers What They Want

You have six to 30 seconds to convince a reader that your resume warrants a complete read, an investment on their end of five to six minutes.

A recent poll I conducted among fellow recruiters revealed that most spend less than 15 seconds on the first page of your resume. Most, in fact, never get past the e-mail note or cover letter (let alone your carefully worded "Objective") and, frankly, human resource managers are no better.

No one has time to waste waiting for a job seeker to get to the point...so the first rule of resume writing is to construct your resume so readers get the information they need fast. A little advance planning is called for.

Be Relevant

Presumably the reader is looking to fill a job in which you are interested, so show how your experience fits the requirements. Don't assume people can or will "read between the lines"—they don't have the time. It's not their job and they don't care about you—yet.

Target Your Reader

You need to understand who your reader is, because different people read resumes looking for different things.

- Recruiters look for "hot" marketable skills because they want to make money marketing you. If your skill set is not in high demand, they won't call unless you are an *exact* fit for a job order they have.
- HR folks look for an exact skill fit with a job first, then your stability, then your personality type.
- Hiring managers look for skill sets first, then how flexible you are, and finally your ability to learn on the job.

Keep It Crisp

People are visual. They like to look at documents that are clean, neat, and well-constructed.

Use Bullets

Sentences, that is. Short sentences are easier to write and easier to read. They also give the reader a sense of action and energy. The reader can get the gist of your experience quickly. You can elaborate at the interview.

Highlight Your Strengths

Whichever strengths (accomplishments) are the most relevant to your reader—they go first. *Always* lead with your best foot forward.

Demonstrate Results

Use ###, %%%, and $$$ to emphasize your accomplishments. One million dollars is less likely to be noticed than $1,000,000. Numbers and symbols jump off the page.

Give It "POP"

Power verbs such as those in Appendix C give your resume "pop," that crisp delivery of the facts. They're high energy and factual. Just rewriting your resume alone with these words will increase your chances of being interviewed by 50 percent.

Be Concise

Your resume should not contain one more word than it needs *to make your point*. Okay?

Connect the Dots for the Reader

Make it easy for the reader to see your fit to his or her job. Before you write your resume, research newspapers, job boards, and Internet ads for positions that are similar to the ones you'll be seeking.

Ensure that the latest "buzzwords" are prevalent. Common key words and phrases (such as JAVA, Audit Trail, channel management, or DWDM) should map to the bullets in your resume.

Scientists and senior executives should prepare an appendix of publications and papers as well. Technical people need a separate Technical Summary page like the one on page 32 for easy identification of your skills.

Include Dates

Always include the start and stop date for every position. If you don't do it, it implies you have something to hide, whether you really do or not.

If you have had many positions in the same company, lump them together in the headers and expand each as a separate line item. You don't want it to look as though you jumped from job to job when in fact you were a "superstar" who had five promotions in three years at the same company.

For example:

> *1981–1999 ABC Telecommunications, Atlantic*
> *1998–1999 VP, Product Management*
> *1997–1998 Director, Access Product*
> *1995–1997 Manager, Broadband*

I'm sure you get the idea. Show your career advancement. Don't leave it to chance, assuming readers will see it on their own.

Be Honest

If you exaggerate or lie, you will be caught. Due to major changes in the law, employers have more rights than ever before when conducting reference checks. If your interview answers don't line up with your references' answers, it's likely that you won't be hired—and you'll never know why you didn't get the job.

Be Humble

It's okay to be a legend in your own mind. Just don't let it leak out onto your resume. Let the facts of your accomplishments do your bragging for you. Wherever you see the word *I*, delete it or replace it with an action word. "I led" would become "led," "I sold" would become "sold," and so on.

You'll get hired for the enormity of your brain, not your ego. When in doubt, have a close friend or confidant read your resume over for any hints of arrogance or pomposity.

Don't Copy Anyone Else's Sample Resume

I could easily add a thousand pages of example resumes to this book, *but* then you might be tempted to copy them, and that's pointless. Everyone can tell when someone hasn't done his or her own resume. There is a "common" feel these days to most of the "pulp" we all see. Your resume should be unique.

Instead, I am going to explain the logic and mechanics behind building your own by pulling mine apart. My personal resume is a value-based resume. I designed this template out of necessity more than 12 years ago to satisfy my own requirements. I continue to recommend it to friends and candidates, because it suits the needs of most readers—especially headhunters. It mirrors the specific logic headhunters use when they receive an unsolicited resume: Headhunters need to know instantly if the job seeker can be a candidate.

1. Is he immediately useful for a project I'm working on?
2. Is she worth my time to meet so I can possibly place her in the future?
3. Should I be actively marketing him to prospective employers?

People who have used this style have had remarkable results. Here's a quick history on how I developed it.

History of the Value-Based Resume Format

In early 1990, I was soliciting business from the president of a major high-tech company to recruit his new VP of sales. Almost as soon as we were introduced, he took one look at my baby face and asked me, "What could you possible have accomplished in your life?" I used to get this a lot. I was 30. I had done plenty, but I needed to demonstrate it. I needed to make up for my seeming lack of experience with energy and drive. He asked me for my bio, and I didn't have one. That same evening I constructed my first value-based resume. You can see that resume in Appendix D.

Normally the employer would be concerned with my "job hopping" from Royal Trust to Harrington to ConsulPro, *but* the "promoted five times in three years" statement redeems me somewhat and gets him to read the Major Accomplishments section, where I intend to demonstrate why I am the *best* person to hire for his project. There is no reason for him not to read the next page.

The "pop" from the first page of the resume sets the pace and tone for the two pages that follow. My prospective client can see that I'm an experienced human resources consultant with initiative, drive, and tenacity. I have the innate qualities he's looking for (remember the New Value Table on page 27).

Major Accomplishments Section

Here I group my accomplishments (instead of skills as a functional resume does) according to the job I'm interested in. In my case, I wanted a consulting assignment to find his vice president of sales. I constructed this section to flow in the order in which he would review my background.

I needed to demonstrate that my consulting experience qualified me to recruit his vice president (hence, consulting is first and business development is second), coax candidates to the table (people management), talk to them on their level about their career (leadership and management of assignments), and finally close the deal (hence, sales is listed last).

Finally, it wasn't really good enough for me to show that I had what he needed, because there were handfuls of executive search professionals in town that could do a fine job. I needed to demonstrate I was *innovative* and *hungry* and *proven* (hence, the marketing and public relations section). The rest of my resume was designed to substantiate my leadership experience and innovative flair. Remember, to him I was a "baby-faced kid."

After I submitted this resume, he called me back for a proper interview, where I showed him my portfolio and laid out the search strategy I had designed based on the research I had completed on his company (more about how you can do this later).

I won the project and I received a very generous job offer as well, which I politely turned down. If I had used a traditional functional or chronological resume, I never would have made it past the first six-second read—let alone to an interrogation (it wasn't an interview). With the plethora of job seekers today, you face the same issue. You may be the best programmer, or sales or marketing person in the world, but if you can't make your resume stand out, you won't get an interview. It's that simple.

Whether you choose a chronological, functional, or value-based resume, try to make yours different! Fresh! New!

In today's rapidly changing, visually stimulating environment, you face an uphill battle trying to get anyone to pick up and read a piece of paper. Advertisers bombard us with 1,500 "messages" every day on television, on the radio, and in the newspapers. Your resume is just another intrusion into someone's busy day, but you can use technology to capture the curiosity of your prospect and deliver a unique presentation of your skills and abilities at little or no cost.

Some Useful Alternatives to Conventional Resumes

◆ Liven up a paper resume. Illustrate your accomplishments with a few charts or graphs in the margins, or even include a separate appendix to the main resume.

◆ Boost your resume with an electronic slide show. With products such as PowerPoint or Astound, you can easily produce a multimedia presentation complete with color, graphics, special effects, and audio. There is no better way to make an impression and deliver a large quantity of information effectively. It's better than being there, because you deliver a controlled message and showcase your skills more effectively.

◆ Use hypertext to design a presentation. This can allow employers to review your background from different angles (for example, specific projects you've worked on, accomplishments, core competencies, and education). It is even possible to distribute it over the Internet.

"You, Inc.": Personal Branding and the Intangible Value of Being You

Knowledge Value

Let's start with the music industry, where personal branding is a well-honed craft. As an extreme example, entertainer David Bowie floated a personal bond issue recently. He offered investors a portion of his future royalties from previously recorded material and receipts from future concerts. The Bowie Bonds were gone within an hour of the offer, for more than $50 million.

In an even more striking case, when Dreamworks SKG went public, investors immediately drove the value of its bonds to $2 billion. Dreamworks was a film studio without a studio, a film, or even a star. All it had was the intangible value of its founders: Steven Spielberg, Jeffrey Katzenberg, and David Getten.

The intangible value of being—that's what the new knowledge economy is all about. Veteran information age guru Stan Davis confirms some insights into the increasing value of people in today's economy in *Blur*:

> A person's "value" is just a measure of how much someone is willing to pay to obtain something from them.

In *Blur*, Stan Davis and Christopher Meyer make the point that the boundaries between your work life and your home life are disappearing. In fact, today the rate of change and the depth of connectivity are so fast that every person, product, service, and company are blurring together.

Computerization and communications have made us all a linked community. There are, for example, nine times more computer processors in our products than in our computers, nine billion CPUs in items such as phones, hotel keys, consumer electronics, day planners, and cars.

As products are more driven by software, they become easier to link together. Intelligence and information become the key *values* being offered in a consumable product (some 90 percent of the value of a new car is estimated to be in the computers and software it uses). And *you* are the value-adder.

Instead of resources or land, "capital" today means human capital. I personally despise the term but it is widely accepted and used by fashionable consultants.

It doesn't take a shoe factory to go into the shoe business these days. Nor do you need raw materials or fleets of trucks. Nike became a shoe industry leader by concentrating on the value-producing capacity of its employees, for design, marketing, and distribution know-how. The real capital is intangible: your knowledge level, combined with an aptitude for application.

Today, employees in the high-technology world especially tend to think of themselves as "free agents"—likened to a professional athlete who is always in training. Knowledge workers are continuously investing in the next set of skills and training, driving up their personal "stock price." This puts knowledge value in the driver's seat. Your brand is unique. How do you give it a dollar value?

Knowledge + Aptitude = Brand

Today we buy the value implied by our favorite brands, and employers do the same! Do you buy generic beer? Clothes? Cars? Not likely.

Personal Branding

Why personal branding is critical for you today:

- ◆ Employers are looking for results.
- ◆ Results demonstrate your qualities, which satisfy their value requirements.
- ◆ Employers are not buying generic beer.
- ◆ Employers will buy the intangible qualities implied by your brand (you are Nike, too).

How do you create a brand—without a million-dollar budget?

It's pretty simple actually. Personal branding is all about making yourself stand out so that people trust you and are interested in you.

To do this you borrow your previous employers brand (names, slogans, and logos) to create an identity that is memorable and desirable to the people you want to reach.

For your cover letter this means name-dropping (which projects you worked on or which clients you sold to). Be specific. Be detailed. Sell the sizzle *and* the steak.

For your resume, it may mean taking the logos (with permission, of course) of the companies for which you worked or products you developed and putting them on your resume for extra punch. Nothing will get an employer's attention faster than a well-known brand's logo, especially if it's a competitor or a coveted account for the sales group.

For telephone queries it's all in how you set the stage. When you are following up your letter, e-mail, or resume try standing out from the other thousand applicants. For example:

Typical:

- ◆ "I sent you my resume."
- ◆ "I'm following up the resume I sent."
- ◆ "I understand you may be looking for..."
- ◆ "Do you need a...?"

Those are the standard opening statements people use when they call me or leave me a voice-mail message.

You, Inc:

> *Mr. Jones, I was in New York when the World Trade Towers came down. My company sent me there a few days earlier to close a deal and told me not to come back without it because there'd be no company to come back to.*
>
> *Mr. Jones, are you getting all the new business you deserve, or are you of the opinion there's room for improvement between now and the end of the next quarter?*

This is not exactly rocket science. It just takes a little forethought and planning to leave a message or start a conversation that is more likely to get you a quick return call.

Your Resume: An eXtreme Makeover

Are you currently in a situation where you are facing an enormous amount of competition for a limited number of jobs? If so, you may want to rethink your resume design. After all, it's your resume that gets you noticed. Getting it read is the first crucial step to securing an interview.

As I have said before, most resumes are boring, and if you put yourself in the shoes of the poor reader wading through all those dead trees you could almost feel sorry for him or her.

Out of pure necessity (read: Many of my friends were calling because they needed jobs), I designed an eye-catching one-page format guaranteed to get noticed *and* read. Essentially it's designed as a marketing document aimed squarely at the reader. It pulls the information that is most relevant to the reader on to one page. It is easy on the eyes and a pleasure to read.

You can use it as a stand-alone resume or, as I am suggesting, a teaser to generate interest and secure an interview where you bring along a full-blown resume. Either way the purpose is to get your phone to ring.

Warning

Although I have not had this confirmed by anyone yet, I think that this type of resume format will only work if you are sending it to a senior executive.

If you send this type of resume to a director, low-level manager, or someone in the human resources department it will hit the trash immediately, because you will be perceived as a threat or too independent.

The new format has seven major areas:

1. Job Objective or Summary.
2. Select Accomplishments.
3. Special Skills.
4. Career History.
5. Education.
6. Awards and Patents.
7. Proof Section.

Let me explain the purpose behind each section. (You can see samples in Appendix N.)

Job Objective or Summary

The first line of the resume needs to tell them what you can do for them. "What's in it for them." Don't even give them a chance to guess, because they might get it wrong. You have a choice of two titles for this section; your choice depends entirely on the situation.

Job Objective

Use this title when you are responding to a specific job from an ad. Your job objective should relate specifically to the ad. Remember the old "you need—I've got"? In this case it's, "You are looking for—I am." For example, which of these two summaries is better?

- A responsible and challenging position that will utilize my past experience, expand my knowledge, and offer opportunities for professional growth.
- To increase shareholder value through correct accounting practices.

If you chose the first one, go back to the beginning and start the book over again because number two is the one which speaks to an employer's needs.

Summary

Use Summary as a title when there is no specific job advertised but, through research (or a good guess), you have concluded they can use your talents.

Select Accomplishments

We use the word *select* on purpose because we want to imply that there are many more accomplishments you could have written down. Bullet point your top three to five accomplishments that you believe would interest your reader the most in relation to the type of job you are seeking. Make sure the accomplishments focus on your strengths and the reader's needs. Use really short sentences with action verbs. Make sure you fill the statements with $$$ and %%% signs to draw attention. Symbols are always easily picked out amidst plain text.

Special Skills

Once again we use the term *special* because it implies you that have more *and* that these are unique to their situation. Following is a list of "special skills" phrases that I have used over and over again with great success. Note that I keep it brief and make statements or claim competencies without going into great detail.

- Audit—keen eye for catching discrepancies.
- Budget maximization—for effective use of finite resources.
- Creative concepts—branding, e-mail marketing, partnerships (VAR/OEM), road shows, strategic partners, seminars.
- Creative concepts—branding, e-mail marketing, partnerships.
- Conceptualization—preparing and presenting deals at the CEO/CFO/CIO level.
- Customer service—reduced claim errors and improved reimbursement times.
- Execution—regularly delivering to fixed schedules against all odds.
- Experimentation—relentless probing for new R&D and product approaches.
- Experimentation—relentless probing for novel solutions and approaches.
- Experimentation—relentless probing for new sales/partnership approaches.
- Expressive clarity—strategic development plans.
- Expressive clarity—industry commentary, outreach, Web writing.
- Leadership—of teams ranging from six to 30 people across multi-country sites.
- Management—optimizing people and finances to meet objectives.
- Management—influence and optimize people to meet objectives.
- Negotiations—national telecommunication issues inside and outside government.
- Sales—track record of leading teams to successful outcomes.
- Organization—canvass volunteers, build and motivate teams, direct outcomes.
- Project management—evaluate programs, initiate projects, develop and execute strategies.
- Project management—working with state-of-the-art high impact systems.
- ROI/Shareholder equity—have built companies using only retained earnings.
- Strategic alliances—technical liaison throughout project life cycles.
- Strategic alliances—to capture market share faster.
- Technical audit—keen sense for system-wide troubleshooting.
- Training—delivered new employee indoctrination.

Career History

Here you showcase your pedigree one line at a time for each employer. As well, you need to write a very tight (two lines max) description of your company, because not everyone will be familiar with your employer. Are you in a well-established blue-chip company or a start-up? It makes a huge difference in how your experience is perceived while your resume is read. Your pedigree matters, especially to headhunters and senior executives.

If you have trouble with this, enter your employer's name in Google and hit the search button. You should immediately see how the company's marketing department has described the company in the first two or three lines that follow the URL title.

Only allocate one line per company and position. If you have had multiple positions within the same company then you want to break them out, one per line, to a maximum of probably five (just enough to show that you're a mover and a shaker). Always make sure this is done in reverse chronological order.

Education

List all you have post-high school. One per line, as the examples in this book show.

Awards and Patents

Again less is more. Use one line to say something along the lines of "23 patents in DWDSM" or "Presidents Club...7 times."

Proof Section

Please don't actually use that as a title. I can hear you laughing, but one of my friends actually did this last year when I was not as specific as I probably should have been. This section runs down the left hand-side of your paper. This is where the concept of borrowing your employers "brand" to build your own personal brand comes in. What goes in here? What would make your reader take notice of you? Could it be your training at another company? Perhaps you have sold into a coveted account? Did you design or launch a major product the employer would be familiar with? There are a thousand things you could put in this section. Let me suggest that you put in only five—make them those five the reader is most likely to be interested in. Here is a list of possible suggestions:

Your Position Sought	Your Reader's Interest	Suggested Graphics
Sales	To whom have you sold? Are there any major accounts you know the company would like to have or would recognize as difficult to get that would make you look like a "superstar"?	Logos of the companies for which you have worked or the major customers you have sold. Perhaps a product you sold if it's more recognizable than the company's logo.
Engineering	For whom have you worked? What major product where you part of designing?	Logos of your employers or customers. A logo or snapshot of the product you designed.
Marketing	What brands have you helped create? Where have you gotten press coverage for your products? What trade shows have you worked?	Logos of your employers. Logos of the newspapers or magazines in which you have had coverage. Media quotes for which you were responsible.
Finance	Have you done an IPO on NASDAQ? Have you secured funding from a major venture capital firm?	Logos of your employers or significant partners with whom you have negotiated.
Administration		Logos of your employers.

Example 1

Turn to Appendix E (page 152) and read the resume of John Walton (not his real name). Can you highlight his top accomplishments and the number-one thing John should emphasize on his one-page resume.

Picking out John's accomplishments is difficult because there are many. The troubling point of his layout is that it hides his biggest accomplishment: winning the president's award not once (which is rare enough), but three times. John is brilliant. He's a solid gold, get-things-done kind of guy, but that's buried at the bottom of page three. I must admit that I didn't even catch it until his fourth phone call to me when he said, "Listen, David I'm the type of guy you want to market because I've won the president's award three times." To which I said, "Well, why didn't you put that on you resume?"

In actuality he had, but I hadn't read past page one. As so many other talented people do, he assumed I would read the entire thing and be duly impressed with his humility when I stumbled to the last paragraph and read of his achievements. He didn't pass the six-second test. He lost my interest in the first few seconds. Too bad for both of us, really.

Now take a look at page 155 to see how we changed it. It certainly has more "pull." If I received this in the mail or by e-mail, I would definitely read it.

Example 2

Turn to Appendix E and read the second resume of Kathy Star. Kathy's challenge was completely different. For years she had been the marketing genius behind several well-known products, *but* her bosses had kept her hidden in the background. She was determined to land a CEO's or President's job at an early-stage or a newly funded software company in her next career move. The trouble was, her resume read like a doctoral thesis.

Check the Xtreme Makeover at on page 151. What we did was accentuate her leadership skills, product marketing skills, and media contacts. We leveraged the fact that every young company needs media exposure to grow by putting the logos of the products she had launched and the magazines and newspapers her products were featured in down the left hand column. In the "Special Skills" section we also highlighted the fact that she was a do-it-yourself type who wasn't stuck in an ivory tower and did not need a whole cadre of staff to execute her ideas.

Other Ways to Dress Up a Resume

Like it or not, we live in a very visual MTV society where people would rather be shown something than have to read it. So what can you do if you are not comfortable being as "out there" as the eXtreme Makeovers? Plenty. The key is to set yourself apart by what my marketing friends call the "hallow effect," wherein the user, in this case you, borrow the credibility of your employers brand. If that's not possible, then even a simple graph or table depicting an accomplishment on the first page of your resume will have an impact.

The graph on the following page was dropped into the resume of a vice president of marketing to demonstrate the dramatic impact his work had on the visibility of his company.

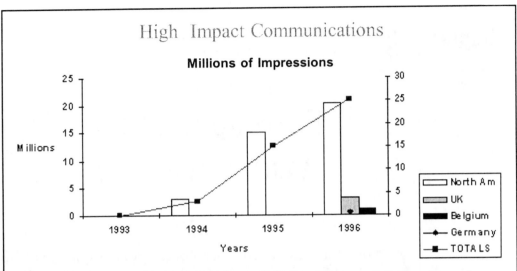

Barry Harry creates compelling, crisp communications that get results for advanced technology enterprises. He has award-winning experience in all the arts of persuasion, including inventive opinion-shaping print and media campaigns on major products and public issues, for some of the world's top companies.

His most recent successes include:

* Igniting a new product from zero to market leader.
* Sparking a media article every day.
* Raising annual media impressions to 25 million from 0.
* Creating a digitally focused selling machine.
* Tailoring high-yield CD and mail campaigns.

Chapter 3

Correspondence With Examples

C over letters are powerful tools that can greatly enhance your odds of getting an interview when they accompany your resume. Often they make the difference between getting an interview or not.

In today's economy, you need to include a cover letter with your resume if you want to be taken seriously. Your letter must be specific to the job, person, and company to which you are applying. Generalized "Dear Sir/Madam" letters get trashed.

Because your cover letter explains your reason for contacting someone and sending him or her a resume, it serves as your introduction of *you*. It highlights your unique qualifications and demonstrates your writing ability and communication skills. For these reasons, *you must write your own cover letter.*

The objective of this chapter is for you to be able to produce a series of attention-getting letters unique to you, not the standard fare copied from a book. After all, your reader wants to know who *you* are and why *you* want to work there, so your words must be your own.

Fortunately there is a fairly straightforward means to compose these types of letters, which I am going to explain. An added bonus for heeding my advice is that almost everyone else looking for a job will use a standard stolen letter from one of the sensational/sexy/standard cover letter, books that are so popular these days. Be original. Most interviewers will have read the book you copied your sensational sexy cover letter from, too. Don't be branded a "me too" candidate with a carbon copy letter—or your resume may be discarded immediately.

There are several different types of letters you need to compose, all of which serve different purposes. Your letters will be used:

- To accompany your resume when responding to a specific job ad.
- To initiate an interview.
- To initiate networking.
- As follow-ups after an interview.
- As thank yous to colleagues for referrals.

Basic Rules

All your letters will follow some general rules.

1. **Keep it short**—no more than one typed page.
2. **Address it to a specific individual**—letters addressed to Sir/Madame, To Whom it May Concern, Friend, or Recruiter generally get tossed in the garbage. Either find out to whom to address it or save the stamp. Get the person's name from the company Website or by calling reception and asking. If the receptionist won't tell you (it happens) call back at lunch and ask the temp. If that doesn't work ask for accounts payable and say, "Oh I'm sorry I have being misdirected. Can you tell me who is (title) and his or her extension number please and then pass me back to reception?"
3. **Highlight all your experience** that is relevant to the job/or company to which you are applying. When you just can't fit it all in, remember Rule #1: Your letter is a "snapshot" or "teaser," not an autobiography.
4. **Include contact information**—put your phone number and e-mail address below your signature block.

You will be writing the following letters, in the order of their importance:

- Broadcast letter.
- Narrowcast letter.
- Covering letter.
- Agency/Recruiter/Search firm letter.
- Networking letter.
- Thank you letter.

Broadcast Letter

A broadcast letter is a direct mail piece and is prepared by extracting facts from your resume. It is sent without a resume to the person in the organization who has the power to hire you. Generally, this person holds a position one or two levels above the job you seek. Unless you are seeking a job in the personnel department, do not send your letter to the personnel manager.

Group your mailing lists geographically so all letters to one area go out at the same time. This makes it easier for you later when you set up interviews and appointments.

The opening paragraph should be an attention-getter, a benefit statement that is relevant to your reader (for example, "Would you like to sell your products to 75 of the top Fortune

1000 companies?"). Then briefly outline your specific reason for writing. If you market yourself as a jack-of-all-trades, you run the risk of turning off an employer.

Next, list some specific examples of your experience and accomplishments to establish your credibility.

Close off with a call to action (for example, "I am certain I could make a contribution to your organization. I will call you in a few days to see if a short meeting is in order."). Yes, I understand you are giving them the option of saying no, *but* if your letter implies that they'd be crazy not to meet you, they won't say no. It also implies that you're too busy to waste your time (or theirs) if there is no interest. This works. Too many other people already use the old "Would Wednesday at 10 or Tuesday at 3 be better for you?" approach. It's old. It doesn't work.

Sample broadcast letters can be found on pages 61 and 62.

Narrowcast Letter

A narrowcast letter is used most frequently when you have in-depth information about a company's needs but it hasn't advertised a position. It is not accompanied by a resume.

Again, it should be addressed to the individual who has the power to hire you, normally your potential supervisor's supervisor. It is easier to have your resume kicked down to a hiring manager than to get it kicked up. Avoid the personnel department. (You can "cc" personnel a copy as a courtesy if guilt compels you.)

An example of a narrowcast letter sent in response to an advertisement can be found on page 63.

Covering Letter

Covering letters are used to answer advertisements, inform third party recruiters of your availability, and/or as direct mail. They differ from broadcast and narrowcast letters only in that they "cover" or are "accompanied" by a resume. A sample letter can be found on page 64.

Agency/Recruiter/Search Firm Letter

Recruiter letters are used to "blast" out your qualifications to all the recruiters and head-hunters that you can find in the hopes they may have an assignment you fit. You always send a resume with this type of letter because busy recruiters will be annoyed if they have to chase you to get one. See page 65 for an example.

Networking Letter

Networking letters are used to inform your immediate circle of friends, colleagues, and trusted advisors (such as your lawyer or accountant) that you are actively looking for new opportunities. You send a resume with it and give your permission to pass it on to anyone he or she sees fit. An example is on page 66.

Thank You Letters

Thank you letters should be sent after all job interviews. They are an excellent way to re-state your interest in the job/company, mention any skills you forgot to discuss in the inter-view, or overcome any objections that you were unable to deal with during the interview.

Years ago when I first got into the recruiting business, my mentor, Bob Henault (The Recruiters Group in Ottawa, Canada), taught me a very simple technique. In the thank you letter, reiterate what you can bring to the table in a sort of "you need, I've got," "you need, I've got," "you need, I've got," rhythm, addressing the major points the employer is looking for. This was simple powerful advice that I have never forgotten. See page 67 for an example.

Sample: Broadcast Letter 1

Address
Town, State
Postal Code

Residence   Business

Date

Private and Confidential

C. Person
Vice President, Sales
ABC Software
123 Smith Street
SmithVille, Virginia

Dear Mr. Person:

Are you seeking a Chief Financial Officer with expertise in financial reporting, budget preparation and staff supervision? I am a results—oriented leader with keen business acumen who executes strategies that attain profit objectives:

• Drove $200K loss to $100K profit in 24 months; positioned company for sale.
• Delivered $300K annual savings by implementing customer/market profitability analyses.
• Increased cash flow $520K annually via implementation of "pay-by-phone" system.

I would be pleased to discuss further details of my background and experience with you in a personal interview.

Yours truly,

Signature

Sample: Broadcast Letter 2

Address
Town, State
Postal Code

Residence   Telephone Numbers   Business

Date

Private and Confidential

D. Person
Vice President, Sales
ABC Software
123 Smith Street
SmithVille, washington

Dear Mr. Person:

I am interested in applying the 15 years of experience I gained as a Sales Manager with OXY Systems to your operation. My background includes all aspects of sales and sales management.

Prior to OXY, I was employed as an Account Executive with ABC Software and as Sales Manager with Old Guys Software.

I have played the role of a hunter and proactively sold Information Security Consulting Services to all medium to large enterprises by dealing directly with the "C" level executives *and* formed Strategic Alliances with a few major organizations that could benefit by bundling/reselling our services with their core offerings.

Although my wife and I enjoy living in the Boston area, we are looking forward to relocating to New York.

I am planning to be in New York during the week of _____. I would appreciate the opportunity to speak to you in a few days to arrange a meeting at your convenience.

Yours truly,

Signature

Sample: Narrowcast Letter

Telephone   Address
Home   City, State
Office   Postal Code

Date

E. Person
Vice President, Sales
ABC Software
55 Franklin Drive
Vancouver, British Columbia

Dear Mr. Person:

Do you need a high-energy, award-winning Software Sales/Marketing and General Manager with a strong analytical capability, an aptitude for strategic and operational planning, and the leadership ability to drive crisp execution?

I have experience in:

- ◆ B2B eCommerceTechnology
- ◆ First/Second Level Management
- ◆ P&L, Revenue/Cost Center Management
- ◆ First/Second Round Start-ups
- ◆ Competitive Research and Positioning
- ◆ Organization and Staff Development
- ◆ Team Building
- ◆ Strategic Planning

I will call you shortly to see if ABC needs someone with my qualifications and whether we should meet.

Yours truly,

Signature

Sample: Covering Letter

Telephone   Address
Residence   Town, State
Business   Postal Code

Private and Confidential

F. Person
Vice President, Sales
ABC Software
55 Franklin Drive
Chicago, Illinois

Dear Mr. Person:

I have 10 years' progressive experience in a healthcare organization. As the meeting planner, I am responsible for the planning and execution of logistics for medical conferences, as well as producing all the publications for the conference.

Throughout my career, I have been overseeing production of publications, vendor selection, facility selection, contract negotiations, food and beverage selections, and arrangements for audio-visual service, transportation, etc. I also conduct reconciliation and make reports on activity costs.

My resume is attached for your review. I would welcome the opportunity to discuss my experience and the contribution I can make to your organization.

I will call you in a week to arrange a meeting at your convenience.

Yours truly,

Signature
Attachment

Sample: Recruiter Letter

Address
Town, State
Postal Code

G. Person
Vice President
Headhunters R' Us
55 Franklin Drive
New York, New York 07417

Dear Mr. Person:

Enclosed is a copy of my resume for review against your client assignments.

I am seeking a position with a progressive company where my sales and sales management can impact shareholder value.

I have worked for Older Guys Software for the last three years. Some of my achievements included:

- ◆ Increased division revenues from $2.5 million to $9 million within two years. Achieved 350% annual profit growth.
- ◆ Expanded division from one to seven states and increased billable staff from 100 to 350+ employees.
- ◆ Negotiated strategic alliances with two Fortune 10 corporations to outsource their administrative and support services organizations.
- ◆ Negotiated multimillion dollar, multi-year contracts.

Previously, I worked for two Blue Chip Consulting companies in business development.

I would be pleased to discuss my qualifications with you in a personal interview. I will call you in ten days to arrange a meeting at your convenience.

Yours truly,

Signature
Enclosure

Sample: Networking Letter

Private and Confidential

I. Person
Vice President, Sales
ABC Software
55 Franklin Drive
New York, New York

Subject: Keeping you updated

Ivan:

I have officially left eStepping Corporation now. eStepping was a provider of end-to-end information security solutions ranging from Professional Services, and Managed Security Services, to Secure Network (Data) Storage. I'm leaving on good terms and with a continued admiration for the technology and marketplace. It's been a great opportunity and a very exciting project and business to build.

- Played the role of a hunter and pro-actively sold Information Security Consulting Services to all medium to large enterprises by dealing directly with the "C" level executives.
- Formed Strategic Alliances with a few major organizations that could benefit by bundling/re-selling our services with their core offerings.
- Generated revenues of approximately $100K per month through direct sales as well as sales via the alliances.

I will be looking seriously for alternatives now (consulting in business practices or full-time assignments) and would like you to have a copy of my resume for reference.

I seek a significant leadership role where excellent management practices, systems, procedures, ethics, a whole lot of common sense, and hard work are required.

Sincere regards,

A. Person
E-mail address

Sample: Thank You Letter

Address
Town, State
Postal Code
Date

Private and Confidential

H. Person
Vice President, Sales
ABC Software
55 Franklin Drive
Dallas, Texas

Dear Mr. Person:

Since our meeting yesterday, I have had an opportunity to review the notes I made during the interview, the requirements of the job, and my own capabilities. Seeing these laid out before me confirms what I previously felt: I can make a very smooth transition and rapid contribution to your operations in the capacity of _____.

I am probably unique in my qualifications for this role. As . . .

+ The position of _____ will require
  _____. I have _____.
+ Your new _____ must have _____.
  I have _____.

Mr. Smith, I enjoyed meeting with you and am very interested in this position.

Yours truly,

Signature
Phone number

The problem with all the previous letters is that everyone uses the same format. Don't misunderstand me: There is nothing wrong with that as long as your skills are far superior to those of everyone else who's applying. In case you are like me (a little on the ordinary side) you may want to know how to stand out so you'll get interviewed before your competitors.

How to Make an Ordinary Letter SHOUT Out Loud

The previous standard cover letters had three components:

◆ **Paragraph 1** indicated which job you were applying for and how you heard about it.

◆ **Paragraph 2** highlighted how your education and experience related to the position.

◆ **Paragraph 3** expressed your interest meeting to discuss the job opportunity.

At least that must be what most books teach people to do because I get 50 to 100 letters similar to this every week. Need I remind you that even if an employer is desperately looking to fill a job they're not interested in *you*—only what you can do for them. Nothing personal.

Ordinary letters are the hallmark of ordinary people. Are you ordinary? I didn't think so. Here's how you can write a value-based achievement driven letters. You need to add the following:

◆ A headline statement.

◆ A P.S. line.

Headline Statement

Your headline statement goes at the top of your letter and should not exceed 20 to 30 words. Your headline comes from your list of achievements. Don't be concerned if it is also in your resume (it should be) just ensure that it's relevant to your reader and the job you want.

P.S. Line

Sales and direct marketing professionals have been using P.S. lines with great success for years, so why shouldn't you? Your P.S. line should indicate when you are going to call. The date and time must be specific *and* you must place the call at the time you designate in the letter. If you are late, the effect is lost and you are better off having never called.

As for the body of the letter, follow the standard three-paragraph format. Just don't beg for the job.

5 Components of a Value-Based Letter

1. **Opening:** Your biggest/most powerful accomplishment.
2. **Paragraph 1:** Indicate you can probably repeat the performance for them. Tell them where you read about the position.
3. **Paragraph 2:** Draw attention to your experience that is relevant to their needs (in bullet form).
4. **Paragraph 3:** Express your interest in the job and suggest that a meeting is in order. Give them the best time to reach you without implying you've got better things to do than to wait for their call.
5. **P.S. Line:** Tell them exactly when you're going to call if you haven't heard from them *and* put the name of his or her secretary in the message (you get this by calling his or her office and asking). This is crucial because secretaries run their bosses' lives at work. They are the bosses' schedulers. In most cases they outlive their bosses, too. They are the corporate memory. They know everything that's

going on *and* they know everyone. I have known more than a few who recommended to their bosses when a subordinate needed changing. Treat them like gold!

Don't copy any of the examples here word for word. Instead, pull your own accomplishments from your resume. This type of letter can be used with any resume format.

Sample: Value-Based Narrowcast Letter

"Increased regional sales from 7,200 to 36,000 new accounts annually"

J. Person

ABC Software

Mr. Person:

I have no idea if I can repeat these results for ABC Software, but I'd like to talk to you about your new OFG software I read about in *eWeek*.

Here's what I've accomplished the last few years with eZ Software:

- Opened, recruited, and managed 7 new sales offices totaling 98 employees (New York, New Jersey, Tampa, Atlanta, Milwaukee, Chicago, and Boston).
- Produced over 500 new accounts with annual revenue over $750,000.
- Sold 1,385 new processing clients in 5 months of operation.
- Increased regional sales from 7,200 to 36,000 new accounts annually.
- Managed 240 sales representatives in 12 Eastern and Midwestern states.
- Implemented training course for new recruits, speeding productivity.

Jim, you're the only person who can tell me if my accomplishments are in line with your expectations for growth.

Perhaps a short phone call would determine your interest in meeting me. The best time to reach me is _____ at _____.

Sincerely,

I. Person
613-555-6995
iperson@abcsys.com

P.S. I will call your office Tuesday November ___ at ___ to follow up. If this is not a good time for you please ask Mary Smith to call me and we'll arrange another time.

Sample: Direct Response Letter

"Dan Person increased sales 530%
at OXY Software"

K. Person
Vice President of Sales
XYZ Systems

Dear Ms. Person:

There is no guarantee that I can repeat these results for ABC Software, but I would like to discuss your requirement for a _____. Your ad in the _____ caught my attention.

Here is what I have been able to accomplish in my career so far:

- Increased sales by 112% at ABC, Inc.
- Made President's Club 10 times at DEF Company.
- Opened 80 new accounts my first year at GHI, Ltd.

Ms. Person, you are the only person who can tell me if my accomplishments are in line with your expectations for this position.

Perhaps a short phone call would determine your interest in meeting me. The best time to reach me is _____ at _____.

Sincerely,

Dan Person
613-555-6995
dperson@abcsys.com

P.S. I will call you on _____ at _____ if I haven't heard from you first or, if another time is better, please have Mary Smith call and we'll work it out with your schedule.

E-mail Letters and Attachments

E-mail deserves a special section because its use as a job-hunting tool is relatively new to most job seekers, meaning that few people use e-mail effectively to get interviews. *All* the rules to which you need to adhere for paper letters also apply to emails.

E-mail has become so pervasive in the last few years that it has developed its own slang. Acronyms such as LOL, IMTTO, and others have no place in e-mails used to further your job search.

The standard cover letters have their e-mail equivalent. Essentially, the purpose of each remains the same but the body of the message changes slightly. First, the most important element is no longer your opening paragraph; it is your subject line. The second change is how you send it as an attachment or in the message.

Let's look at how an ordinary recruiter or executive deals with e-mail resumes.

Unsolicited E-mails

I use Outlook's filters to aggregate unsolicited e-mails in one basket, which I generally glance at once a day—if I have time. Most recruiters use similar filtering software.

As do most people, I only have a limited amount of time each day to deal with interruptions. Unsolicited e-mail is an interruption, but I need to deal with it. Emails get deleted rather quickly in the following order:

Every subject line that contains the text, "Please review my resume. I look forward to hearing from you" gets deleted. Why? First, I've opened and read a ton of these over the last couple of years and I have *never* found one that was worthwhile. Other recruiters and employers may have different experiences, but according to the recruiters' network that I am part of, my experience is typical, and they do the exact same thing. Yes, I may actually miss a golden nugget—but I am willing to risk it.

Second, that subject headline and others like it gives me no compelling reason to look at it. Very selfishly, people believe that I should be honored to get their resume for review. Frankly, that's never going to be the case. You need to show me in your subject line what's in it for *me*.

You need to learn to treat your subject line as a newspaper headline. If the front page of the *Wall Street Journal* were full of headlines that said "please review" or "read this," the *Wall Street Journal* would go out of business. Newspaper headlines need to reach out and grab the reader.

Everyone selectively filters in only those bits of information that are of interest to him or her because we are constantly bombarded every day with thousands of messages vying for our attention. I personally read headlines such as "DOW up 100 points" or "ABC Software expands." What do you scan for?

So returning to the example of my inbox, I have several subject headlines that would be of interest to me if I were looking for a financial person. There are three resumes that might be of interest.

- ♦ Resume—Project Manager—Analyst—Manager.
- ♦ Resume—Investment Banking—Financial Services.
- ♦ Resume—Financial Executive—Controller/CFO.

I would likely open the second one first because of the reference to banking and financial services. I might open the other two, but only if I haven't found what I'm looking for.

So, of the remaining e-mails, can you guess which get read first? Generally all those *not* from Grassis Greener and BlastmyResume. Why? Not just because they're spam, but because they aren't directed to me. They're free for any recruiter who wants them. And the problem there is that a product loses its value when it becomes a mass-market commodity and the subject line is vanilla, so, unless I'm looking for an "engineering manager" or "fashion buyer," I won't read them.

Now you maybe thinking, "Yeah Dave, but if I blast my resume to 10,000 recruiters then I'm bound to find someone who has an open position." It is possible, but here's the catch and the reason you need to target your efforts: Contingency headhunters, who make up the vast majority of Third Party Recruiters (TPRs), compete amongst themselves to fill a client's assignment. It's quite literally a race! Now imagine what happens when you blast your resume to all of them and they all submit your resume as a candidate. You lose. You lose because they'll take you out of contention. Companies aren't in the habit of paying multiple fees to multiple recruiters for referring the same candidate, which is what they would be forced to do in that situation. Are you so good the company would gladly pay for you two or three times?

Rules for E-mail Correspondence

Make sure your subject line gives me a reason to open and read your message. "George Smith's cover letter and resume" isn't a good enough reason for me to open and read it. (I'll give some examples of what *is* shortly.) Your cover e-mail should get my attention within the first three sentences. Remember you only have 10 seconds to make an impression, so don't waste it telling me you're "hardworking and enthusiastic." So is everyone else! Here are some tips:

- Direct your e-mail to a specific person.
- Use an inviting subject line.
- Capture my self-interest with your opening paragraph.
- Tell me how to reach you.
- *Embed your resume into* the e-mail message.

Directed E-mail

Take a look at the figure on the following page. Here's an example of an e-mail that got through to me. Got opened. Got read and got a phone call back. What made the difference? You can't see it by this example, but it was actually addressed to me—David Perry—not "occupant" or the e-mail equivalent hr@anycompany.com.

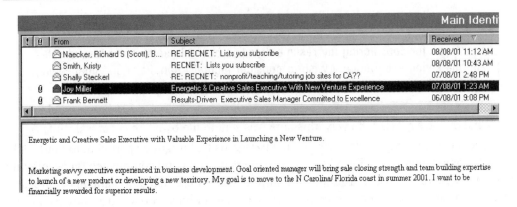

Energetic and Creative Sales Executive with Valuable Experience in Launching a New Venture.

Marketing savvy executive experienced in business development. Goal oriented manager will bring sale closing strength and team building expertise to launch of a new product or developing a new territory. My goal is to move to the N Carolina/ Florida coast in summer 2001. I want to be financially rewarded for superior results.

Inviting Subject Line

Next, the subject line got my attention because she creatively told me something about herself: energetic, executive, venture experience. Here are some other examples of great subject lines:

- ◆ "Latin America business gladiator—Market entry strategies"
- ◆ "Experienced technical sales consultant in New England"
- ◆ "SONET Guru"

Opening Paragraph

The opening paragraph from our example really grabbed me:

> *"Marketing-savvy executive experienced in business development.*
> *Action-oriented manager will bring sale-closing strength and team-*
> *building expertise to launch a new product or develop a new territory.*
> *My goal is to move to the North Carolina/Florida coast in summer 2001.*
> *I want to be financially rewarded for superior results."*

This one is direct and to the point. If I had a job order, I'd read it immediately. If I did business in southern United States, I'd read it. If I did business with start-ups, I'd read it. I did read it, and I liked what I saw: a successful candidate with real skills and accomplishments who's motivated to accelerate her career and knows were she wants to work. It doesn't get a whole lot better than this for a "blind" e-mail.

Your e-mail needs to identify your unique selling proposition. What's so special about you? As do cover letters and resumes, your e-mail only has seconds to catch my interest.

Here are a couple of good paragraphs:

> In the course of your search assignments you may have a need for an experienced Chartered Accountant for the position of Director of Finance or Corporate Controller in either the private or not-for-profit sector.
>
> I am a financial management executive with 14 years experience in corporate accounting and asset-based lending.

Here is an example of a *bad* paragraph:

> Thank you for considering my resume detailing over 10 years of successful sales and marketing experience. I wish to consider mutually beneficial and challenging opportunities where my listed experience can be further leveraged.

What did he say? (By the way, this note was sent to me by a "new car" salesman.)

Also, take advantage of keywords and buzzwords that might be pertinent to your background. For example, use CPA, CA, risk management, audit and compliance for accounting types. Most recruiters use a program similar to one that I use that automatically aggregates your resume and e-mail, and allows keyword searches for buzzwords and acronyms that are pertinent to the position. Having them in your resume and e-mail makes it all the more likely I'll find you.

Tell Me How to Get Hold of You

I need your e-mail address if for some reason you don't want me just to hit the "reply" button, *and* I need a phone number that has an answering machine or service, because I don't want to be chasing you forever.

Embeded Resume

Always copy your resume into the text of the message below your signature block. You can send an attachment as well, *but* a lot of email systems are now set up to automatically reject text attachments because of the viruses they may carry. *Don't chance it!*

Examples

Take a few minutes and pick out for yourself what's wrong with these e-mails.

Subject: Senior Level Leader Seeks Career Advancement Opportunities

> To Whom It May Concern:
> I am seeking a career opportunity with a firm whose primary objective is to build business solutions to meet internal and external customer expectations. I am also very interested in facilitating the development and implementation of integrated operations that cross divisional boundaries and lead to technological leverage.

Bad subject line. Bad first paragraph. No "what's in it for me".

> Subject: Marketing Superstar With Diverse Background Seeks Dream Job
>
> Dear Recruiters and Hiring Managers,
> I am a marketing public relations expert with 14 years' experience developing strategic plans for consumer products, trade, and entertainment sectors. My outstanding background well suits the needs of marketing, public relations, and corporate communications departments within consumer product-based corporations, studios, and networks. I am seeking a senior level executive position with a Los Angeles–area company.

Good subject line and okay first paragraph. Even though this one is not directed to me in name it still got opened.

Here's a perfect example:

Subject: Applicant for CIO or Operations Management Positions

Dear Recruiter:

Are you looking for me...a chief information officer or operations executive who can make a substantial and immediate contribution to your client's productivity, cost structures, and product/service quality and can strengthen its financial results?

I offer a unique value proposition that I'd like to put to work for one of your clients: While with the "Big 5," I enjoyed considerable success working with industry leaders such as Ford, HP, P&G, MetLife, Microsoft, Levi Strauss, and The Gap on both domestic and global fronts.

I am confident that my skills in revenue and profit contribution are easily transferable. If you are working with a client that would benefit from a unique combination of technology and operational leadership, call me.

I am open to relocation. In recent years, my compensation has been in the range of $200,000 to 270,000. Resume follows and is attached. Thank You.

Here's an example of what *never* to do:

Subject: Worth considering!

Dear Friend:

With a proven issues/solutions approach to business—and the ability to communicate with the top floor or the shop floor—I feel it could benefit both of us for you to know more of my background. I have heard about your excellent search work from a colleague of mine, a well-connected colleague, who also mentioned now would be a particularly good time to contact you. He heard you have a search in progress for an exceptional sales/marketing executive where I would be a strong candidate. It's unlikely you would find me through normal channels, so I'm writing to you instead.

Using referrals as a door-opening technique is a great idea. But, the referral has to be real. This guy is so far out to lunch it is obscene. He's really trying to impress me. What's more, the e-mail mentioned a colleague, "a well-connected colleague." Who? And if that were so, he'd know my name and not call me "friend." What are "normal channels"? Guess I'm too incompetent to find him. Luckily he contacted me first. Unfortunately, he forgot to hide the e-mail addresses of the other 217 recruiters to whom he also sent it. No cigar!

Here's a good e-mail, though there are too many "I"s.

Subject: Client of Ken Zaborniak—Communications/Marketing Management

David,
I received your name from Ken Zaborniak, who suggested that I get in touch with you regarding opportunities in the Ottawa area.

I am seeking a position in communications/marketing management where I can utilize my skills in strategic planning, project management, writing, and client relations. I have also worked in media relations for a number of organizations.
I will contact you within the next week to touch base and see what the market looks like.

Best regards,

Here's another *bad* e-mail:

Subject: Dave Johns' Resume

Enclosed please find my resume for your consideration. I am actively seeking employment for a senior management position in sales, marketing, or general management. Listed below are some search guidelines that may help in your placement efforts.

Current compensation:
$110,000 base salary, 20% bonus potential, 6% profit share, 5% retirement fund, 401k with 4% matching funds
4 weeks paid vacation plus 10 holidays
Severance agreement
Complete medical/dental co-pay benefits
Relocation package

Targeted compensation:
Minimum $115,000 depending on the position and location
30% compensation package with bonus, profit share, or retirement
401k with matching funds
4 weeks paid vacation plus scheduled holidays
Severance agreement for 12 months
Complete medical/dental benefits

Anyone who gets this specific in his or her demands is kidding himself or herself. I'm glad we found out what he wanted in terms of compensation before we had to go to the trouble of finding out what he did.

> Subject: Experienced Director with Proven Track Record Seeking New Challenges
>
> To Whom It May Concern:
> My purpose in contacting you is to present your company with a candidate who has consistently met and exceeded revenue goals and performance objectives. What I do best, is to build and develop sales teams and organizations using focused leadership and process management skills that celebrate personal and departmental successes.

Good subject line. Bad opening (because it's not addressed to a specific person). Good opening paragraph.

◆◆◆

I hope these critiques are enough to show you what constitutes a good e-mail message that will get opened and what doesn't. Remember to put your resume into the body of the e-mail in Rich Text Format as well as attach it as a Word document. Don't use other text packages. Everyone uses Word to send resumes, so use Word to ensure that your message can be read.

Direct Marketing with E-mail

Let's tear a page from a direct mail advertising expert. To even get "opened," your presentation and content must speak to a star's interests by providing information that answers this question: What is in it for me? To be effective, your e-mail content must to be short and concise, while providing just enough detail to move them to contact you. You're writing an ad for you!

These steps will help you write and successfully advertise yourself:

1. **Your headline or title promises something to your prospect.** You must grab their attention at the outset by way of your headline. Consider the following headlines: "Visual C++ ARCHITECT!" or "CRM Guru for Chicago." Obviously, the "guru" headline is correct. The first headline is just like dozens already in my "out" file resume basket. The "!" symbol was picked up by my spam filter and automatically deleted.

2. **A good lead follows a good headline.** A good headline will pull in inquiries. Now write a good lead based on that. The key here is to elaborate on your headline with facts. Again, taking our example, don't waste a paragraph detailing your specs. Get at their Value Table ASAP. Example: "Managed two new releases that beat the competition and garnered 35% market share." Wow—a techie who understands time to market issues? Where's the phone?

3. **Be convincing by providing ample information about your accomplishments.** Ask yourself: What can I say or write here to convince an employer to pick up the phone and call? Remember that grabbing their attention in your headline and lead is winning only half the battle. The next step is for you to convince them to call or e-mail for more details. Bullet three to five accomplishments that might be relevant to their market.

4. **Learn to storyboard.** If you want dramatic results, learn how to write your copy as a docudrama, where you have an introduction, a conflict, a climax, and finally a solution. Organize your points according to importance. For example, try this one that certainly got my attention: "My company bought me a one-way-ticket to Brazil and said I couldn't come back without the deal. They had no money for my return trip! That was three years

ago. I was successful and now I am ready for bigger challenges. Mr. X, do you hire people who think on their feet?" Do I!

5. **Executives don't read boring e-mails.** If you must, get a marketing friend involved (better yet, a copywriter). Insist on having a unified message developed in conjunction with your resume. Link back to your Website and relevant news articles. One effective way for you to connect is for your e-mail to speak directly to them. Appeal to their emotions, their needs, and their desires. If possible, add testimonials. Never make the mistake of talking more about yourself and your success than what is in it for them.

6. **Be believable.** Tell the truth. You must develop trust quickly. If you have to choose between being clever or ingenious and being honest and truthful, go with the latter. Employers will be able to discern for themselves that you are being honest. Deliver what you promise because employers talk and that's how you develop your network. If they can't use you perhaps someone else in the company or a friend at another firm can.

7. **Tell them how to contact you.** It's important to put an e-mail address, a phone number, or even an address as the final step. Make sure you cover the basics. I know this sounds lame, but you'd be really surprised how many people forget the basics.

8. **Test, test, test!** Write your ad different ways and test it until you see which one pulls.

9. **Leverage technology.** Whenever you can, use technology to personalize each e-mail. Word, Excel, and Outlook are a great power trio.

The Perfect E-mail Campaign

It was serendipitous that I received this e-mail from George Pytlik just days before my book deadline. George's resume had been automatically filed in my unread resume file a few days earlier and on a break I quickly scanned the file looking for nuggets, and I hit the mother lode. As the true professional he is George let me reproduce this for you. Following is what had arrived:

Naturally after an introduction such as this, I had to know more, so I immediately picked up the phone and asked him about the concept and meaning behind "Brand Width."

George told me that, even when they do understand the need to market themselves, most people mistakenly think of individual forms of media in isolation from each other. They decide to send an e-mail, a letter, or a direct-mail piece, never giving thought to how these things work together to create a much more powerful, bigger-than-life impression of your brand.

According to George, "You are the brand when looking for employment. It's no different than marketing any other product or service. You need brand width or your message loses its impact along the way."

For George's campaign he ensured that his business cards, e-mail messages, Website, and direct-response pieces all send the same message. They look similar. They use the same color scheme, images, fonts, and concept. This way his message has a much greater chance to burn through the clutter of our over-communicated world and actually get noticed.

I asked George to describe the process he used. He explained that his e-mail messages are sent using a sophisticated tool (there are a number on the market, including shareware products that do this) that allows him to send to a database of prospects, sending each one a customized message with slightly different wording. It pulls specific fields out of the database to customize the message in the e-mail. Even the desired job classifications can be modified slightly depending on who receives the message, so that the request is as targeted as possible. If he wanted, he could even use the person's name in the message or subject line. The subject line is also modified depending on what's in the database information.

George added that, in our spam-filled world, e-mail messaging is getting much more complex. As a result, the subject line has become more important than it ever was, and the wrong subject—or even the wrong word in the subject—will get your message trashed faster than the blink of an eye. According to George, there's no point in making your subject clever or vague. It should be clear and to the point or it won't be read. Other issues will help ensure your e-mail actually arrives at its destination. Try to send only to the party intended, not to the secretary's desk or another layer below your prospect's desk. Two potential problems exist if it arrives somewhere else. First, there's a good chance it won't ever get sent further on. Second, even if it is forwarded, there's a chance that your message (unless it's text-only) will get scrambled during the forwarding process and will look a mess when it arrives.

Here are some other guidelines to keep in mind courtesy of George Pytlik:

◆ Make sure your e-mail works on both HTML-capable and text-only e-mail clients. This uses a process called a "multi-part" message. The message itself contains two versions of your e-mail. When it arrives, the software tells the message to either display the graphical version or the text-only version.

◆ If your e-mail is a graphical one, make sure your e-mail contains a link to a Web-based version of the document so people using e-mail clients that only partially support HTML messages can click the link to see the message the way it was intended.

◆ Make sure your message contains a link to your Website that's clear and direct. And make sure your Website works properly on all browsers. Don't assume that if it looks okay in yours it will look the same way on other people's computers. What if your most promising employer just happens to use a Web browser that displays the information differently? Sometimes links don't work properly or graphics display in unexpected ways. Don't take chances.

- Make sure your e-mail and Web page both contain links directly to your resume in other formats. I use both Adobe Reader (PDF) format and Microsoft Word (.doc) because these are pretty universal. The benefit of the Word document, though less graphical, is that it can be copied and pasted more easily and can be imported into many internal systems, usually not supporting the PDF format. The benefit of the PDF format is that it can look stunning because you have a full range of graphical capabilities and it will print exactly the same way it looks even if the recipient doesn't have the fonts you used on his or her system.

Always remember that, when creating your Word version, use only fonts that exist by default on standard Windows operating systems all the way back to Windows 95, and forget the use of graphics. Save those for the Website or risk the dreaded *#$@**&%@#$.

To see how George followed through on the promise of his email I visited his Website at *www.adwiz.biz* shown here.

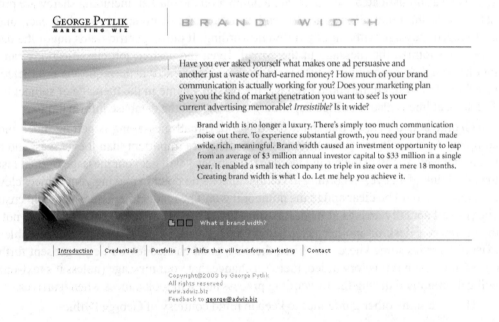

This is a perfect example of how to do it correctly, and I am grateful he allowed me to show you. I clicked on the link to his site and found a completely integrated Website that showcased his skills, abilities, and, more importantly, his philosophy toward marketing. I am definitely interested in helping him look for a new opportunity. What recruiter wouldn't be?

If you decide to attempt this you need to understand that it is critical that yours look as professional as George's. Nothing says "amateur" faster than one of those free Websites that are included with Microsoft Word. Your site has to project a professional image, or all your work is for naught. You have to follow through on your pitch.

In the event you're like me (graphically challenged), surf over to this site and customize one of its templates: *www.templatemonster.com*. Although you will have to write the text yourself, buying one of the templates will give you a great head start. Visitors automatically assume that if it looks great, it must be great. I have used its templates myself twice now.

Chapter 4

Hunt Your Own Head
Advanced Job Search Strategies

You've heard it before: "Finding a job is a full-time job." It's more true than ever today. If you think you can just produce a great resume and wait for a recruiter to call, think again. The $160-billion staffing industry was the first victim of the current economic slow-down. Job orders, the lifeblood of the business for most firms, dried up in February 2001. Many headhunters will not be busy until the second quarter after an economic recovery. Most recruiters have never seen a recession and have no idea how to work through it, so you need to do a lot of the heavy lifting (formerly done by recruiters yourself).

Most people will limit their search to the traditional approaches, such as answering want ads, surfing job boards, and asking friends. That's a great start, but everyone else is doing that too. Now your "friends and colleagues" are probably competing against you. You need to outthink and outwork them. Working smarter requires rethinking some of the old job search sources. Here's how.

Working Smarter

Here are some new twists on traditional approaches:

Newspapers

Although many companies have switched *part* of their recruiting budgets from newspapers to online classifieds, major dailies are still a rich source of job openings. Most major papers have their own online classified section in which jobs are archived, often for 30 to 60 days. Go back through those in your area. You would be surprised how many jobs do not get filled the first time they are advertised.

- Sign up for the online classifieds "job alert" programs, which notify you of matches with your background.

- *Always* check the classifieds. Display ads (or the "Career Section," as it is commonly referred to) are very expensive, as much as 100 times more costly than the classified "word ads." Most companies switch from display career advertising to classifieds during belt-tightening exercises. So you must read both or risk loosing out on 90 percent of the ads. Recruiters use word ads almost exclusively because anyone actively looking for a job reads both the Career Section *and* the Classifieds.

- Review the "Appointments" section for the names of recently promoted or appointed executives in companies that may be of interest to you, and send those people your resume with a broadcast or narrowcast letter.

- When searching through the Classifieds, remember that companies always try to hire locally first for expense reasons, so don't rush out and buy papers from another city unless you are prepared to move there and pick up your own expenses. On the other hand, if an out-of-town company is recruiting in your local paper, you can bet it's desperate, so jump on them quickly.

- Here is a link to the largest daily newspapers in America in order of their circulation: *www.bizmove.com/media_directory/top_100_newspapers.htm.*

Most importantly read headlines like a bloodhound, looking for clues on who's expanding or contracting who could use your expertise. Local news is always a great source of ideas for new job opportunities. Be the first to contact a new company in town or offer to help a company that's in trouble.

Job Boards

Most job boards allow applicants to search for openings *and* post their resumes. Job boards do not share information with each other, so you need to register with as many as you can find time for. Only those companies that pay for access can post a job or review your resume.

The best-known job board is *Monsterboard.com*. With its huge advertising budget, Monster is everywhere. Its "brand" is second to none. The largest, purely Canadian site is *Workopolis.com.* Online advertising is still very expensive, though, and some sites are so expensive that they are really only used by the Fortune 1000.

Newspaper giant Knight Ridder, which operates more than 90 major newspapers across America, also owns the Website Careerbuilder. Careerbuilder is gearing up to challenge the dominant position held by Monsterboard. Because it seemlessly integrates print and Web advertising—the best of both worlds—Careerbuilder is in a good position to challenge Monsterboard.

If you are looking for a job with a start-up don't just look on the big boards. Small and medium-sized companies (where all the growth—and ironically, in my opinion, where all the job security—is) use lesser-known niche boards.

- Gartner Group predicted that 95 percent of all employers would be advertising online by 2002. Looks like they were correct. The top job boards are *Monsterboard.com, Yahoo.com, Dice.com,* and *Careerbuilder.com.*

- Even though *Monsterboard.com* is the largest, it has only 75,000 corporate customers out of a possible 10 million just in the United States, which is less than 1 percent of the U.S. market. At last count, Interbiz found there were 25,000 job boards on average grossing $1 million per year. At the moment there is no way to register for multiple job boards through a single registration site or portal.

- Register yourself at all the top job boards and you will cover 2 percent of the market. The usual rule in marketing is that the top 20 percent of any type of company own 80 percent of the market. Not true here, obviously. So, who owns the 98 percent? No one. You find the other 98 percent other ways. Keep reading.

- The most important thing to remember about job boards is the rule of "last in first up." The way job board works is fairly simple. For the benefit of recruiters (who pay the bills), searches for candidates are delivered in reverse chronological order according to when candidates enter their data. So if you really want to have yours appear in the first 25 retrieved, you need to go in every day and update or change something in your file and refresh it. You are fooling the system into believing you are newly on the market, which is exactly what recruiters want to know. It's a simple trick with big benefits.

- Register at as many of the niche-"free" sites as possible and cover another 1 percent.

- Register at local job boards in your area, because most employers advertise and source candidates locally first.

- Most job boards will allow you to set up "agents" to compare your profile to new jobs and inform you when they come along.

- If you do not want to be bothered at home by recruiters, you should list an e-mail address as your main point of contact (preferably one you can cancel when you find a job).

- If you are deeply concerned with privacy, you should not give out your home address or home phone number.

- You should also de-list yourself when you are off the market.

Following are some of the other job boards that carry a lot of technology positions:

Sales

www.rxcareercenter.com

www.ama.org/jobs
jobs4sales.com
marketingjobs.com
marketingmanager.com
www.AccountManager.com
www.SalesEngineer.com
www.salesjobs.com

www.sales-jobs.net
www.topsalespositions.com

Engineering

www.asme.org/jobs
www.ieee.org/membership/benefits/careers
www.internec.net/jobsearch/mechanical.html

www.systemsengineer.com

www.mechanicalengineer.com

www.electricalengineer.com

www.engcen.com

www.engineer500.com

www.engineerpro.com

www.engineering-direct.com/main.html

engineerjobs.com

civilengineeringjobs.com

www.telecomcareers.net

RF Engineers

www.telecomcareers.net

www.rfjn.com

UNIX

www.unixadminsearch.com

www.uguru.com

www.careermarketplace.com

www.justcomputerjobs.com

www.computerjobs.com

www.ugu.com

webusers.anet-stl.com/~cheng/unixjob.htm

www.unixreview.com

Telecom

www.telcorock.com/JobBank

www.telecom-jobs.net/employers.cfm -

www.telecomengineer.com

 www.telecomjobsite.com/telecomjobsite/ jobs.html

www.engineeremployment.com

www.engineeringjobs.com

www.engcen.com

Call Centers

www.collectionjobsite.com

www.CallCenterCAREERS.com

www.callcenterops.com

www.callcentermagazine.com

www.TMCnet.com

www.TelecomClick.com

Note: For those engineers in the audience you should read these boards, too If sales are growing, it's likely the engineering department is as well.

Newsgroups

I believe newsgroups are the best source of jobs. There are more jobs on newsgroups than anywhere else now, primarily because they're free for everyone—for employers, for recruiters and for you. Headhunters use newsgroups 50:1 in favor of job boards. It was impossible to keyword-search newsgroups until *www.workinsight.com* was invented. WorkInsight, of which I am a part-owner, aggregates jobs from all the technology newsgroups in one place.

Technology User Groups

User groups are much like newsgroups and facilitate the exchange of ideas between people with common interests. They are an ideal place to ask for job leads because they are industry-focused. The easiest way to find user groups of interest to you is to go to Google Groups (formerly Deja) at *groups.google.com*.

User groups exist on every conceivable subject. Type your favorite subject into Google, and you'll likely find dozens—if not hundreds—of groups dedicated to that subject.

Referrals From Friends and Relatives

Personal referrals are an excellent way to find a job. You need to involve *all* your friends and colleagues in your job search. Most companies post jobs internally before going to newspapers or third-party recruiters. It's not just because it's cheaper, but because the employer has someone on staff who is interested in and ethically accountable for the new person he or she refers.

Ensure that your network of friends has your resume in electronic format and permission to forward your resume to hiring managers on your behalf. Ask them to let you know about it.

When referred by a colleague, always ask the referrer how you should follow up. Some people will want you to call; others won't. You need to abide by their wishes or they will not refer you again.

When you are looking for a job, your lawyer, doctor, dentist, colleagues, and accountant are part of your network, especially if they all work in your community and know other companies in your line of work.

The best way to highlight your recent availability is to use a networking letter. (See Chapter 3 for an example.)

Networking: Only Her Hairdresser Knows for Sure

As a new friend of mine recently discovered that networking is one of the fastest ways to find a new career opportunity in today's job market.

The goal of networking is not to call someone to ask for a job, but rather to speak with individuals who, because of their place in the business community, may be in a position to offer career advice, guidance, direction, and introductions.

In Pavla Selepova's case it was her hairdresser who initiated the chain of calls and meetings and yet more calls and more meetings that ultimately led to the creation of a new position tailored to her talents and interests.

> *As the story goes Pavla was talking to her hairdresser, a trusted source of advice for years, about her need for a financial planner. After their meeting, the financial planner referred Pavla to an HR consultant who met her based solely on the planner's recommendation...who referred her to a friend...who in turn insisted she had to meet her next-door neighbor. "I was so impressed with Pavla's energy and commitment I created a job for her," said Lynda Partner, president of GotMarketing.*

For Pavla it took only 34 days to become the vice president of customer relationships. She did have one advantage that you might not: She was working with David Craig a managing partner of Drake Beam Morin. DBM is one of the world's largest outplacement and career transition firms, where Craig teaches people how to network because, if positioned properly, the "rejection" rate is very low, as most professional people want to see themselves as helping others in need. Who doesn't want to help someone else?

David compares finding a job to something most techies are already familiar with: the launching of a new product. "First, one needs to have a clear understanding of the product (oneself), its value-add, and be able to articulate this in convincing terms. Second, one needs to conduct market research to determine, through investigation, what groups (companies, industries) are out there in the marketplace," says Craig. Then you match both sides up.

◆◆◆

Here are some tips to help you network effectively:

- If you haven't already done so, make a list of everyone you know and send a networking letter telling them you will be calling for their advice.

- Starting with the people you know best forces you to make those calls. It gets easier with each passing call. Don't jump right into your agenda, start off by asking, "Is this is a good time to talk *and* what's new?" They'll get around to asking about you soon enough.

- Your goal is to get referrals. So how do you ask in a manner that won't put people off? Generally it's better to be subtle, so rather than, "Can you give me the names of all your friends?" try, "Who else should I be talking to?" This is far less self-serving.

- Dig for information about industry trends or trends in your functional area or specialty. Listen for plans for new products or services. Seek out emerging markets, hidden jobs, and companies that are hiring. Focus on anything change-related. Change equates to opportunity.

- Your telephone calls may soon produce face-to-face meetings. Set the stage by asking how much time they have because these meetings are courtesy calls that must reflect well on the referee or they won't continue to help. State your purpose clearly and directly. Share your excitement and enthusiasm, and ask for advice and ideas. In general, you listen more than you talk.

If you're wondering what to ask, try these openers:

- How does my resume look?
- What would you change or modify?
- Do you have any advice or ideas for me?
- Who else should I be talking to?
- Are there any groups or organizations I should attend?
- Are there any books or publications I should read?

These are my two personal favorites:

- What would you do if you were me?
- To whom would you be talking?

Your contact network should always be growing, and the best way to expand it is to seek out new people and build relationships. It doesn't really matter whom you choose, so long as you like them, they like you, and you can help each other. And when you get a job let them know they helped with a quick note of thanks.

So what was the most important lesson Pavla learned? She says, "You'd better be at your best when you talk to them or they will not connect you to their network. While you look for a job, every conversation you ever have is a potential job interview and should be treated as such, even with your hairdresser." Or, as David Craig says, "Network or...no work."

The Hidden Job Market

Not everyone gets outplacement help and advice, so you need other tools, too. Newspaper ads, job boards, newsgroups and friends will only lead you to 30 percent of all the job openings available at any given time. To reach the "hidden job market," you need to take a more proactive approach. You need to learn how to direct-source job leads the same way a headhunter does.

How to Direct Source Your Next Job

For most people, the Internet is a mess. The Net suffers from too much information and too little structure. Most people looking for a job using the Net quickly get overwhelmed with the quantity of responses for a search engine. Your job is to use the Net to find the jobs you identified in the first chapter.

Headhunters use three basic means to find new job openings if they are going to actively market a candidate and find them a position:

- ◆ Targeting competitors.
- ◆ Obtaining referrals through associations.
- ◆ Conducting Internet searches.

Targeting Competitors

The easiest place for a recruiter to sell your skills is to a direct competitor or at least someone who is in your industry. The easiest way to research an industry is to use Hoover's (*www.hoovers.com*). It has a tool called "Industry SNAP SHOTS" where you can quickly get the lowdown on who's who. The Industry Master list, which covers more than 300 industries broken into 28 sectors can be found at *www.hoovers.com/feature/block/detail/ 0,2417,18_1757,00.html*. If you need it, Hoover's probably has it—*and* it's free.

If you want to work for a direct competitor or need to prepare a list of competitors for an upcoming interview, go back to *www.hoovers.com*. Enter in the name of the company and hit the "capsule tab." This will not only give you a snapshot of the company but also its competitors. You can play the competitor's competitors game all day at Hoovers and never finish. This is a very rich resource. There are 53 major categories for technology workers.

There are two other key organizations you can also try: *www.edgar-online.com* and *www.herring.com*. Both are also very good for company information. To find out about companies to avoid try *www.fuckedcompany.com*. Be prepared, though, because the comments know no bounds and are not for the faint of heart.

Associations

The next best way to research an industry is through its associations, and the best site to help you find the association most related to your interests is the American Society of Association Executives (*http://info.asaent.org/Gateway/OnlineAssocS/list.html*). The site carries Canadian listings, too. For a complete list, see Appendix N.

Remember that it is the job of the president of each association to help members, and this includes referring candidates. They are often the first to know when a company is about to expand.

Internet Searches

By far the best way to discover new opportunities is by doing structured search engine queries, which is fairly easy to do. Here's how to do targeted research, step by step.

Step #1: Develop a target list of companies

Remember Exercise 1 where you decided where you wanted to work? That's the baseline query for our search.

When you do targeted research, generally you concentrate on an industry or a geographic preference (in this example, Dallas). You can use whatever city you like.

Suppose you need to find the names of all the photonics companies in Dallas. There are easy ways to do this using the Net. For a cheat sheet on how to use search engine syntax to find a job please see Appendix L. For our example, go to *Google.com* and try to replicate what you learn here. (Your results will be different from mine because Web pages will have changed since I did this example, but the principles are the same.)

Using the advanced search tool type in "optical Dallas" in the box marked "with all of the words." Then go down to the third box marked "with at least one of the words," and type in "directory conference."

We are using this search string to instruct Google to return the lists of Websites that have a directory *or* conference that is related to optical *and* Dallas.

The first hit is for the largest society for optical engineers. The second brings back the Converge digest Website that has a directory of all the optical companies in Texas, and therefore Dallas, with links to their Websites.

Converge also has a list of virtually every photonics company in the world *and* has an up-to-date listing of new photonics companies, including which companies have recently received venture capital. This is a great resource for finding out about new jobs before they hit the newspaper. These sites will tell you which companies are where. Make a list of all those in your area.

Going back to the SPIE site, you'll also find that it has its own job board, which is much more focused and more relevant than the large catchall job boards.

Step #2: Find people who can hire you

Once you have a target list of companies with whcih to work, you need to find out who the people are in those companies that can actually hire you. A headhunter would likely pick up the telephone and ask a series of mind-numbing, thought-provoking questions that would deliver all that golden information to him or her. You might not be so inclined, so here's another way to do it.

Go to each company's Website and gather names. If you're lucky, every Website will provide the complete identification of all its senior executives, including names, titles, phone numbers, career summaries, and sometimes e-mail and photos! Web information should be up-to-the-minute accurate, but if you have any doubts, make a phone call to confirm it.

> *In our example, I'm looking for a sales job so I'll find the VP of Sales, VP Sales & Marketing, VP Marketing or General Manager.*

If you're having difficulty finding the name on the site, go back to Google's advanced search box and type in the company name in the first box and these words in the third box: vice president sales marketing director. You don't need the brackets and do not put in any commas or punctuation.

Still nothing? Try this. Go to Google and type in "*@company.com." The asterisk (*) is a wildcard for everyone who has that e-mail address. In other words if you're looking for someone from xyz company and you know that Bob Smith's e-mail is bobsmith@xyz.com try searching "*@xyz.com." You should get a lot of people if it's a big company. Also try searching using the Groups tab on Google. There is no guarantee you'll find who you're looking for, but you may get lucky.

If you need to know how the e-mail is formatted, search the company you are interested in for its news section, and find an e-mail address on one of its press releases. That should do it.

The search string in our example will bring you:

◆ All the people who are, or have ever been, vice presidents *or* president *or* directors of sales and/or marketing for that company.

◆ The resumes of a whole pile of people from that company who you may be able to phone to coax information from them.

Once you have the name of the individual who is one rung up the ladder from the job you want, you need to process his or her name through Google again. This time you put the person's first and last name in the first box and the company name in the third box.

This will produce a list of press releases and news articles in which the person is mentioned, as well as conferences he or she has attended. Read an article or two and clip something memorable to use in your Narrowcast letter.

When you send them the letter, you'll be able to say, "I read your article in *X Publication* about *topic* , which prompted me to write." Very powerful.

Other sources of information on who can hire you can be obtained by referring to annual reports, 10(k)s, and proxies. You can look up the phone numbers in Standard and Poors or another large general directory, or call toll-free information: (800) 555–1212.

Annual reports provide valuable organizational information, division and subsidiary data, locations, names, titles, revenues, numbers of employees, discussions about strategy and growth plans, and sometimes even photos of employees.

My personal favorites are 10(k) reports, which are required by law and disclose names and titles of senior management, each executive's number of years with the company and a career summary, and his or her age. Age is relevant because shareholders have a right to know when key managers might be approaching retirement, which could materially affect the performance of the company—and you may be who's on their mind as they do some quick succession planning. Often, 10(k)s provide plant locations and delineate a company's lines of businesses. They must also state if anything could adversely affect the company's performance or stock price, such as a major lawsuit or pending environmental expenses.

Proxy statements are required to disclose the compensation paid to the four highest-paid executives. Proxies also provide detailed background information on the board of directors.

You can obtain a hard copy of the annual report, 10(k), and proxy for free by calling the company. Most companies post these reports on their Websites in Adobe PDF format.

Exercise

Conduct a search on Google for Coldfusion jobs.

Step #1: Develop a target list of companies

Decide on key words that are specific to the type of job you are looking for. We'll use Coldfusion and JavaScript. The generic words you'll need are *job, resume, submit*, and *free*. We are using this search string to instruct Google to return the lists of Websites that have Coldfusion jobs but are not ads for resume submitting businesses.

By using the "without the words" option, we also instruct Google to exclude those sites selling mass resume-submitting services, which would just clutter our results. Why type in the word *free*? Because the word *free* is not used on corporate Websites, nor is it used in job descriptions. It is, however, used to sell resume-submitting services.

Choosing the advanced search option and entering the words one by one brings you the following results:

Words	Number of Hits	Relevance
Coldfusion	512,000	Low
Resume	16,300	Low, includes candidate resumes
Submit	2210	Low, includes job boards
Job	1190	High, all Coldfusion jobs
JavaScript	658	Very high, Coldfusion and JavaScript jobs
Not "free"	483	Very, very high, excludes all the resume submittal sites and shows just the jobs.

Note: If you find you get too many search results outside the geographic area you're interested in, try putting in area codes instead of cities to localize the results. Area codes are a more exact means of honing in on a city. If you were to just do a city search for New York or Dallas or Ottawa you would probably miss 75 percent of all the jobs.

A complete set of instructions for search engines can be found in Appendix L.

Step #2: Find people who can hire you

Once you have a target list of companies to work with, it becomes: "Now I need to figure out who, inside the company, can actually hire me." This isn't all that difficult. As you'll see, the Internet puts most of the information right at your fingertips. You just need to know how to find it.

In my case I'm looking for a programming job so I am likely to be hired by a VP of engineering or VP of marketing.

If you're having difficulty finding the name on the site go back to Google's advanced search box and type in the company name in the first box and "Vice President Engineering Director Marketing" in the third box.

This search string will bring you:

- ◆ All the people who are, or have ever been, vice presidents *or* president *or* directors of engineering or marketing for that company.
- ◆ The resumes of a whole pile of people from that company from whom you may be able to phone to coax information.

Once you have the name of the individual you're looking for, process it through Google again, with the person's name (first and last) in the first box and the company name in the third box. This search should net you a wealth of information (such as press releases or news pieces). If you're lucky, you might get a contact list from a conference he or she attended, which could provide you with a lot more leads to follow up later. The best "find" is an opinion piece or white paper he or she has written or published, which you can read and reference in your narrowcast letter.

Alternatives

If you really get stuck go to Eliyon (*www.eliyon.com*) and sign up for a free subscription. Eliyon is likely to have what you need. I was one of its first beta testers and subscribers. It had profiled some 15 million people at last look.

Other Search Engines

Here are other popular search engines you can use and descriptions of how you should structure your queries. Always use the advanced search page.

AltaVista (*www.altavista.com*)

<keyword(s) go here> AND (job OR jobs OR submit OR opening OR send OR apply OR post OR position OR opportunities OR EEO OR candidate OR candidates)

example:

(software AND sales AND Ottawa) AND (job OR jobs OR submit OR opening OR send OR apply OR post OR position OR opportunities OR EEO OR candidate OR candidates)

AltaVista Canada (*http://ca.altavista.com*)

<keyword(s) go here> AND (resume OR curriculum-vitae) AND education AND experience AND objective

Excite (*http://corp.excite.com*)

Resume AND experience AND education AND <keyword(s) go here> AND (job OR jobs OR submit OR opening OR send OR apply OR post)

Google (*www.google.com*)

<keyword(s) go here> +job +jobs +submit +apply +send +post +opening +position +opportunities

HotBot/NBCi (*http://hotbot.lycos.com*)

SEARCH <keyword(s) go here>
MUST HAVE: submit apply position

MSN (*www.msn.com*)

<keyword(s) go here> AND (job OR jobs OR submit OR apply OR send OR post OR opening OR position OR opportunities)

NorthernLight (*www.northernlight.com*)

<keyword(s) go here> AND (job* OR submit OR opening OR send OR apply OR post OR position OR opportunities OR EEO OR candidate*)

Yahoo (*www.yahoo.com*)

Resume <keyword(s) go here>

Bringing New Opportunities to You

There are a vast number of services you can subscribe to for free that will bring information on hot new companies straight to your desktop every morning. My favorites for technology news are Venture Wire for North America and Silicon Valley North specifically for Canadian news. Here's where to go and sign up for several other industry news sites:

- Semiconductors: *www.semiconductoronline.com*
- Engineering of all types: *www.eetimes.com*
- Venture Wire: *http://professional.venturewire.com/prosub.asp*
- Silicon Valley North (for everything Canadian): *www.siliconvalleynorth.com*
- Optical: *www.lightreading.com/employment*
- Wireless *www.wirelesstoday.com*
- Telecom: *www.telecomweb.com*
- Satellites: *www.satellitetoday.com*
- Cable: *www.cabletoday.com*

And, of course, Google (through Google News) has an immediate news alert on any subject you need.

First Contact

If you're like most people who are looking for a job, you probably hate the whole process. You probably hate it most because you don't like the rejection that comes with writing resumes and phoning to try to get interviews. You're not alone. Some of the best salespeople I know have severe allergic reactions to "cold calling," which is why I'm going to suggest you consider an alternative.

Any well-targeted, well-written letter usually will get you a response, often without the need for a phone call from you. This is especially true when you put the name of the hiring manager's executive assistant in the letter and tell them when you're going to be calling. The assistant may actually be waiting for the call. Most people will call you first. These are, however, very busy times for the technology industry, so here's what I suggest you do to make this even easier:

Send the letter in a plain white envelope with your potential employer's name, address, and so on handwritten on the front. Do be sure to put a return address. Why? Because you don't want your letter to be discarded before it is opened. Slick packages and "labels" are a dead giveaway for resumes; a handwritten envelope is not.

So your letter will get opened and I guarantee you'll be the only one that has the executive assistant's name in your letter.

Next, wait until the time when you said you were going to call, and call. The assistant will get the call. This is what you say:

> *"Hi Mary, this is David Perry calling. I sent a letter to Mr. Person a week ago and promised to follow up at this time. Do you know if he has had time to read it?"*

Most books will tell you to try to go around the secretary or assistant. I'm urging you to involve him or her in the process because nobody—*and I do mean nobody*—knows more about what's going on than assistants do. He or she will tell you whether Mr. Person has read it or not because it's his or her job to know *and* because his or her name was in the letter *and* because you've just used his or her name and asked for his or her advice. The psychology behind this is simple, but very powerful.

If the boss is in, he or she will tell you whether this is a good time to talk or not. If not, he or she will likely schedule a time for you to call back *and* put the call in the boss's schedule and your letter and resume on the boss's desk.

Be prepared to start a relationship over the phone, because more often than not executive assistants will actually get into great long conversations with you or even tell you what type of people the company is hiring *and* why. They do this not because you seem like a great guy, but because you've shown them the respect they deserve, unlike the last 50 schmucks who tried to trick them.

My best advice to you is to work with assistants to get you in front of their bosses. It's easier than trying to go around them—and smarter, too!

Chapter 5

The Active Interview
A Two-Way Situation

Little more than 200 years ago, a leadership revolution of epic pro- portions heralded the birth of a new country. America, fresh with new people, bright ideas, and high hopes, changed the world's ex- pectation of leaders and leadership forever. Knowledge workers today are having a similar impact on technology companies everywhere.

Back then, General Washington created a level playing field by re- cruiting skilled professionals who knew the enemy and what it would take to win. From leaders such as von Steuben to Lafayette, Washington sought individuals who could make a difference in the outcome of the war. They turned a ragtag collection of farmers and shopkeepers into the troops that led the United States of America to victory. Today's business leaders toil on a different battlefield, but it is a battlefield nonetheless.

The Impact of Knowledge Workers

Today managers must lead, workers must innovate, and organizations must capture market share, or they, too, will be relegated to a footnote in the history books. A company's critical assets are the minds of its employ- ees. Novel strategies and tactics are required for the innovative and cre- ative pursuit of clear and common goals.

We live in a world of ideas where the employee who, when asked to participate in a monotonous routine, will contribute little, whereas the same employee engaged in creating the future will produce well beyond the ex- pectations of management and significantly contribute to the creation of wealth.

For employers looking to grow their companies, what has mattered most in the last five years are three questions: Can you make me money? Can you save me money? Can you increase my efficiency enough to cover your costs? Yes? You're hired! With so many talented people now looking for jobs, interviewers believe they have the dual luxury of time and choice on their side: time to make a leisurely decision and the luxury of having many people from whom to choose. Unless you want to get stuck in an endless cycle of interviews, you must change the interviewer's perception of current reality. You must change their process, at least as applied to you.

Demonstrating your net worth by appealing to an employer's value table when constructing your resume and cover letter is a start. It'll get you an interview, but that's just the beginning. Now the real work starts and more subtle factors come into play.

No More Team Players—Please!

Let's face it: You know you're good at what you do, but so are many other unemployed people these days. The bar for new recruits has been raised very high for hiring. Today, many employers prefer to make no decision rather than possibly make a wrong one.

To leap ahead of your competitors you must subtly demonstrate the unwritten qualities that are now *the most* in demand:

- Leadership skills.
- Communication skills.
- A bias towards action.
- Passion.

Leadership Skills

Employers today don't need any more "team players." The last 10 years in the technology industry have shown that team players are often afraid to voice their opinions. Nobody needs another hanger-on! Every technology company right now is battling the clock trying to increase market share in the midst of a cash-crunch and a dwindling capital market.

Employers want leaders at every level of the organization who are capable of galvanizing the talented people they still have around ambitious goals *and* motivating them to succeed. You must convince an employer that you will have a decisive influence on its ability to win and that you are an integral part of its solution. If you can't communicate your personal commitment and drive through your words and actions in the interview, you won't be the first choice.

Communication Skills

Crisp, clear, concise communication is in. Abstruse pontification is out. Save the high-level theorizing for the pub. Your demonstrated ability to direct and motivate staff is the key to interviewing success. Employers hire articulate candidates before *all* others. No one has the luxury of time to interpret what they think you said. Fuzzy thoughts and clumsy speaking skills are not indicative of clear thinking.

Be prepared to get into details with your interviewer. Be prepared to relate your accomplishments to his or her needs. Be prepared to tell the interviewer why he or she has to hire you!

A Bias Toward Action

You must demonstrate your ability to take action with limited or imperfect information. Wall Street brutally punishes companies that don't move quickly to make changes, regardless of whether the company has the correct information to make a decision. Learn how to consult your "gut feel" when nothing else is available. Employees at all levels cannot be afraid to make tough decisions. Product life cycles are now measured in months, not years. The time for debate is short. Lately, investors from around the world have punished acts of indecision with dramatic sell-offs.

You have to be viewed as decisive, because anyone else is viewed as a liability. You demonstrate these traits as much by what you do before the interview as how you conduct yourself during the interview. The easiest way to research an industry is to use Hoover's (*www.hoovers.com*), which has a tool called "industry SNAP SHOTS" that allows you to quickly get the lowdown on who's who.

Passion

Most people coast through life preferring to be safe rather than sorry in their career, but I have had the great fortune to work with brilliant technical people who were also passionate about what they did and who wanted to leave their mark on the world. People such as Juan Guillén, with whom I first worked at Simware and who recently passed away, have a fire in their belly, a zest for life, and a sense of urgency that infects everyone around them. They challenge you to stretch and open your mind to new possibilities—to envision what is *possible*, not what *is*.

For most employers, it really doesn't matter if *all* of your ideas are bright, or even right—just that you have some and that you participate. Passionate debate leads to breakthroughs that create new industries and new wealth for all.

Winning a Job Offer Is Preparation Meeting Opportunity!

The ability to separate out those who excel from those who are just ordinary is the difference between skilled executive search professionals and the one-size-fits-all "bodyshoppers." The recruitment industry is in tatters now, so you have to think and act strategically if you want to position yourself as an exceptional, can't-do-without, contributor.

Start by being passionate about who you are and what you do. The world doesn't need any more "team players." It needs more Juan Guilléns.

Current Reality: Understanding an Employer's View of Risk

A headhunter or corporate recruiter is taking a risk in recommending you to a client. The hiring manager is taking a bigger chance in choosing you (or any candidate for that matter). If he or she makes the wrong choice, at minimum, time and money are wasted. At worst, a bad choice could jeopardize the recruiter or manager's job, perhaps even the success of the company. Team players are vanilla. Are you? Employers want to hire von Steubens and Lafayettes.

Most employers know this but can't recruit for it. Hiring managers, in a company of more than 50 people, rarely interview more than two to three people each year, and interviewing skills get rusty if they're not used regularly. For that reason alone you need to direct them to the relevant parts of your background that explain why you're the best candidate (and you thought the previous chapters were a pile of work). Otherwise you're leaving your career to luck and the odds are stacked against you!

Twenty years in this business has taught me that most employers don't really know what questions to ask in order to examine a candidate's suitability, but this rarely deters them. Anyone can talk, but very few can interview correctly. However, interviewers will, nevertheless, make some very real if not potentially inaccurate observations regarding your candidacy for a job after your meeting. It's your responsibility as the job seeker to show that hiring you is not only the best decision, but a safe one too—even a career booster for the interviewer.

Basically, interviews are artificial situations during which you're trying to make a favorable impression to get an offer; the employer (who is probably more nervous than you are) is trying to assess if you are the best candidate for the job—often in one meeting. Talk about pressure. There's a lot on the line for both of you.

Most employers realize that interviews do not mirror real life and they can fix a bad-hiring decision with a severance check. Fixing a bad career move is harder. Today, successful career due diligence for a knowledge worker means developing a method for assessing during the interview process which companies have a promising future. The following steps will help ensure that you make the impression you want and that you can make the assessment that you need to.

Prepare for the Interview With Extra Vigor!

You will come across in the interview as a stellar can't-do-without candidate, if you research your future employer strategically. This isn't all that difficult. It's just a lot of work. You need to:

- Determine your net worth.
- Map your experience to their needs.
- Rehearse your message.

Determine Your Net Worth

Employers hire to solve problems, so, before you do anything else, you need to determine where the employer's focus is. If you have an ad, read it carefully. Go to the company's Website and look for an expanded version. Next, read its latest press releases and the president's message. Why is the company hiring? What's happening in its market to necessitate this hire? Try looking at the job from the employer's perspective. Determine, based on your accomplishments, what you can contribute. You won't get an offer without doing this work.

The basic question to ask yourself is this: How will hiring ME into this position make this employer money, save it money, and/or increase its efficiency? Essentially—what's in it for them?

Map Your Experience to Their Needs

Analyze your career and be ready to describe your most significant accomplishments, your short- and long-term goals, and your strengths and weaknesses *as they apply to the job you're being interviewed for*. You must have concrete examples ready, so do what my mentor, Bob Henault, taught me to do years ago. Construct a T-account with the employer's needs on one side and your skills/accomplishments on the other. Do it ahead of time.

They Need	I've Got

You should also prepare five to eight "success stories" that you'll draw on to create analogies during the interview. Your success stories must be relevant to the hiring requirements. From the T-account, pick out instances where you used those skills successfully. Write this down in point form first and then write a two-paragraph description around each. The act of writing this on paper forces you to organize your thinking *and* log the details in your mind for easy retrieval later. Review it just before each interview.

After you have your five success stories, you need to rehearse them until you can deliver them effortlessly. In an interview, your ability to retrieve information quickly is important. It builds your confidence. It increases your resolve. And it amazes the interviewer, who will be expecting long periods of silence as unprepared candidates search for answers. Someone once told me that for every two seconds an employer watches the other candidates think, they lose credibility and you get bonus points. Drill yourself hard. Make your own flash cards if need be.

Prepare analogies—sound bites and detailed stories—that will give you a large jump on other candidates. Most people wait until they get into the interview and wing it. When they don't do so well, they try to make up for their forgetfulness in a follow-up thank you letter. This won't work. You, on the other hand, won't be wasting your time trying to recover from a poor interview; you'll be closing the deal.

Remember: An employer is taking a risk hiring you and so is the interviewer. In many cases people are judged by the people they hire, so they want and need to see you as "all together." After the last couple of years of the "So tell me why I should work here—man," attitude from twenty-something kids with no experience and rings in their noses, employers will embrace candidates who do their research. Life has changed. Your approach has to keep up.

Rehearse Your Message

You need to be prepared ahead of time. Many popular books tell you how to find out about the job once you are in the interview so you can tailor your answers to questions. You had better go into the interview with as much corporate information as you can. If you wait until you get there to gather "intelligence" you'll get creamed! The game will be over before it's even just started—especially if you're sitting at home unemployed. Employers expect to be told why you want to work for them early on in an interview. You had better know.

In Chapter 2, you detailed your accomplishments and how they demonstrated your net worth. Now is the time to map your accomplishments to their needs. Go back and refer to those questions using the T-account approach.

Practice, Practice, Practice

Here are some pointers to help you control the interview and keep you on track.

- ◆ Prepare your answers.
- ◆ Be ready to describe your most significant personal and professional accomplishments, your short-term and long-term goals, and your strengths and weaknesses as they apply to the job you are interviewing for.
- ◆ Have concrete examples ready. *Do not* rely on your ability to "ad-lib" in this situation.
- ◆ Rehearse until you can repeat it without thinking. It eliminates choppiness. The ability to retrieve information smoothly and quickly is important.
- ◆ Prepare five or more success stories.
- ◆ In preparing for interviews, make a list of your skills and key assets that apply to the job for which you're interviewing. Then reflect on past jobs and pick out one or two instances when you used those skills as part of an accomplishment.
- ◆ Write down five or six recent accomplishments that demonstrate those skills. Include facts, figures, and dates—everything that you can.

Be specific and practice answering questions using your accomplishments as examples. You need to be able to recall the details in your examples instantly without having to pause and think about it, and when you write something down you log it in your mind.

It's really all about your self-confidence. You know that you know your job but it's always hard to perform at the time of an interview no matter how much experience you have. It's stage fright, plain and simple. Rehearsing helps because once you get enough practice, the answers will begin to flow smoothly, and the interviewer will send you subtle positive buying signs (such as a knowing nod or smile). You'll pick it up and begin to really relax and be yourself. Your interviewing will improve dramatically. Your self-confidence will rise and your body movements will exude confidence too. Mary Kay would say, "Fake it till you make it." You'll win.

Write Your Own Job Order

Recruiters typically write a "job order" for each position they recruit. Essentially, a job order details the requirements for the position including all technical aspects and duties—and more importantly it details the corporate culture ("buttoned down" or "relaxed," "aggressive new player" or "market leader").

You should write out your own job order. It will show the employer you're serious. You do not need to show this to any of the interviewers. It will be evident in the types of questions you ask and the background information you refer to during the interview that you have done your homework. That alone will increase your chances of receiving an offer.

Okay, so now you need to get the information. What do you need to know about the company? Where do you start? Our basic job order form can be found at Appendix G. Granted, this is a lot of work, but you will be demonstrating to the employer that you are part of the top 2 percent of candidates, not the bottom 15 percent.

Most of the corporate information you need to collect can be gathered from an employer's Website; the rest you can get from sites such as *Hoovers.com*.

Other Sources of Information

Ask personal contacts within the company—people with whom you've had information interviews, or friends, or friends of friends who can give you the real skinny on the job. A coffee shop is an economical means of obtaining inside information (no, don't set up a table to SPY—invite someone out for a coffee). Here are some other options:

- ◆ Review their annual report.
- ◆ Use library reference material. The public library in any city is a great resource. Call the librarian and tell him or her what you're doing and he or she will be able to point you to resources you won't find on the Web.
- ◆ Use the Internet. Go to Google and type in the company name and see what comes up. Read some of the latest articles to see where it's headed in the marketplace. Type in their company name and the word *resume*. This will bring you all the people who have worked at this company and/or who have their resumes out on the Internet. You may find the very person you're looking to replace and his or her resume. You may want to e-mail him and find out why he left. You might want to see what job he took in his next company, what his background was, and so on. It will also bring you job ads for the company by which you can see how long it's been looking. You may get a more detailed description of the job.

Sometimes you can even see how much time and effort the company has put into filling the job. If it's on 10 job boards, you know the company is serious. If it's been on those job boards more than six weeks, it may be desperate or very particular about who it hires. In either case, you'd better find out why.

Research 101

Okay, here goes. If you want to research a company here is what to do. This list is in order of my preferences based on the likelihood of success.

Start with the company home page if they have one. To find out go to Google and do a search on the company name. Unfortunately you are likely to get a lot of hits. Look until you find the one for the company Website and start there. Create a file on your computer's desktop and, using the "Save as" command in the "File" menu, save the following in the new folder: company history, products, news (last six months), management roster, and anything else you think is appropriate. Hard disk space is cheap; bandwidth isn't.

Now you may ask, "Why not just read it there?" Why? Because inevitably the day you need to review the information for a job interview, the Webmaster will have taken it down for modifications. It happens all the time.

Remember: The company Website has all the information that it wants you to know. What about stuff it doesn't want you to know, such as competitors? The home page is the easy stuff everyone will do. How do you go deeper? How do you research its competitors and really impress your interviewer? Here are the sites I use to get an insider's view. I have divided them according to whether you are researching a public or private company.

Privately Held Companies

These are the hardest to research cheaply, because the companies don't have a strict requirement to report to anyone but their limited shareholders. So if they have anything to hide (such as pending litigation or poor financials) you really have to dig. Here's where I would start:

- Dun & Bradstreet (*www.dnb.com/us*). D&B are masters of information, much of it a la carte or free.
- Million Dollar Database (*www.dnbmdd.com/mddi*). Information on approximately 1,600,000 U.S. and Canadian leading public and private businesses.
- Forbes 500 Largest Private Companies (*www.forbes.com/businesstech*). Always one of my favorites but only useful for the largest companies.
- Business Week (*www.businessweek.com*).
- Thomas Register (*www4.thomasregister.com*). If you need information on a manufacturer it's probably available here. Also a great place to research who makes competitive products.

Publicly Held Companies

This is a lot easier, but still requires a great deal of work.

- Dun & Bradstreet (*www.dnb.com/us*). Always start here.
- Edgar Online People (*http://people.edgar-online.com/people*) searches SEC Filings by a person's name or displays all people associated with a specific company name. Very useful.

- Million dollar Database (*www.dnbmdd.com/mddi*). Information on approximately 1,600,000 U.S. and Canadian leading public and private businesses.
- Lexis Nexis (*www.lexisnexis.com*). Legal, news, public records and business information.
- Corporate Information (*www.corporateinformation.com/home.asp*). A free site that requires registration.
- Financial Web (*www.financialweb.com*). Stocks, SEC filings
- Fortune 500. The 500 largest U.S. companies.
- GrayMetalBox (*www.graymetalbox.com*). Trade Scenarios Reports and Research Center.
- 123 Jump (*www.financialweb.com*). Get SEC filings, company news, links to home pages, current quotes and graphs, and so forth.
- Wall Street Research Net (*http://stocks.internetnews.com*). The most comprehensive company news, EPS estimates, links to home pages, and so on.

What Should You Be Looking For?

1. News about the industry trends:
   - Where is the company now?
   - Is the company in growth mode or not? Why? What factors affect each? Where is it in that cycle?
2. Financial Information:
   - What do the numbers say? How is the company's balance sheet, income statement, earnings, EPS, dividend(s)? (Remember: Cash is king.) What is the debt-to-equity ratio?
   - How is the stock price doing? Current price vs. historical price? Why is it moving? What or who has the greatest effect on it?
   - How is the stock doing against its competitors? Against the market as a whole? Are there other companies you should be interviewing with, too?
   - What do the analysts think?
3. Strategy:
   - Have they got one (usually in the annual report)? What is the company's short term and long term strategies/objectives?
4. Market Share:
   - Are they dominant players? Why? How big is the market? What size of the market do they own? What's the next market?
   - Innovations/New Products/Patents: Does the company have any new products/services/patents?

4. Domestic vs. international markets:
   - Is the company strong or weak domestically vs. overseas? Where does the company make most of its profit?
   - What do each of the regions and products/divisions contribute to the whole?

5. Technology issues:
   - Is the company technologically driven?
   - How does the Internet affect them? Do they have a strategy to address this very real threat?

6. Legal and regulatory issues:
   - Are there any possible pending bills or regulations that might have a significant impact?

How to Close the Employer *the Day Before* the Interview

When you wrote your job description for this position for which you are about to interview, I hope you found the phone number of the interviewer's executive assistant. Use it to call the employer the day before and say, "Hello, my name is David Perry and tomorrow at 10 a.m. I am scheduled to meet with Mr. X. I just wanted to call today and ensure that his schedule is still intact." Wait until the assistant checks. Thank him or her and hang up.

How many people do you think do that? I've only had it done once to me and from that day on I've suggested it to all my candidates. E-mail works equally well, as long as it is to the assistant. It is too easy for the employer, if he thinks he's going to be running behind, to cancel you. Assistants are assistants because executives can't organize their days. Assistants rarely cancel something they've had to set up themselves. It would be similar to saying, "Hey, I can't do my job." It won't happen.

My advice is to work with the assistant. I guarantee that tomorrow you will be the only person the assistant remembers. You'll be at the front of the line even before the interview because you've got rapport with the most important person in that office: the assistant, who will let the boss know how professional you were in calling ahead.

Exercise: Strengths and Weaknesses

Quickly, list your top 10 strengths. Write them now without thinking too much.

1. _____
2. _____
3. _____
4. _____
5. _____
6. _____
7. _____
8. _____
9. _____
10. _____

Now, put yourself inside a prospective employer's head as you consider your strengths. Would you pay cold hard cash to buy your strengths? Avoid false modesty, but also be brutally honest and realistic with yourself.

Now let's do the fun one: weaknesses. Don't you just love it when you get asked about your weaknesses? I used to hate this question when I was trying to get a job. As if I wasn't nervous enough: interviewing on my lunch hour...changing in the washroom of the closest fast food joint—but I digress. I hated trying to rationalize why I wasn't a detail person.

Let's do this the smart way. When you're assessing your weaknesses, think about what prospective employers might consider the areas you could improve upon, based on your work experience that you mapped to their job. If you identify a weakness, be prepared to explain what steps you are taking to improve. Having a weakness isn't necessarily the end of the world—unless you're in denial, you know it's a serious problem, and you're not doing anything about it.

List five weaknesses now.

1. _____
2. _____
3. _____
4. _____
5. _____

Employers, and in particular their HR folks, will use terms and phrases such as "areas that need improvement," "soft spots," or some other nebulous term to try and understand your weaknesses without putting you on guard. They simply want to understand from an HR perspective the "whole you," and the package (or in some cases baggage) you bring.

Try to think of this exercise as HR trying to understand your maintenance costs. How much money will the company need to spend to train you up or indoctrinate you into their culture? That's what they are trying to get at. These types of soft costs often get factored into a hiring decision, especially when candidates are viewed as being equal in stature. Do not drop your guard.

So what do you do now? First off, it's important to remember that no one is perfect (in spite of what I've told my children) and, in many situations, a perceived weakness can be turned into a benefit. For example, being ruthless may be just the trait you need to save a company or turn around a product line. In a situation such as that, being a "team player" is actually a real negative.

At an interview, your goal is to reveal only those weaknesses that could be perceived as strengths for the job you are interviewing for. Here are a couple of examples:

Weakness: "I've been told I am a little slower then other programmers."
Strength: "However my bugs are near zero, and the QA guys love me."

This is an example of using a weakness as a strength.

Or

Weakness: "People say I'm a perfectionist."
Strength: "BUT I just like to ensure we're 100 percent on the important issues before we press forward."

This is the absolutely perfect answer if you're in accounting, finance, project management, or even contract negotiations, areas in which it's the little issues that often account for large losses when projects go sour.

I'm certain you get the idea. Make sure to assess your weaknesses this way before an interview.

The Last Thing You Do Before the Interview

The best thing you can do after you press your suit, lay out your clothes, and check your shoes, before going to sleep the night before the interview is rehearse the answers to the following questions, which will be asked in one way or another:

- Why do you want this job? (Remember the annual report? What was it the president focused on? Growth industry, revolutionary technology—try to pull something out of that if you yourself don't have a compelling reason.)

- Why do you want to leave your present company? Why did they leave you? (You might, for example, indicate that your company went through 19 rounds of layoffs and you were in the last one *or* you opted for a package the first round to reassess your career.)

- Where do you see yourself in five years? ("Doing your job!" is never a correct response for too many reasons to mention. Pick a target position two levels above the one you are interviewing for—*if* you're very ambitious. If you don't want to move quickly, tell them. That's okay, too.)

- What are your personal goals? (Be careful not to mention that you've started another company on the side and need this one to pay the bills. I always suggest people talk about family or education, which will broaden their perspective in their new position.)

- What are your strengths and weaknesses? (Refer to the last section.)

- What do you like most/least about your current company? (Acceptable answers: My boss is my age and a great guy so if I want to move forward in my career at this point, I need to look elsewhere. Your company would provide... *Or* It's a small family-owned business and my boss is the president's son so....)

Most employers ask these basic questions *and* most candidates can't answer them in context with the job they're being interviewed for. Don't be caught unprepared!

During the Interview

From the moment you drive into a potential employer's parking lot you are on stage. Assume you are being observed from the moment you step out of your car. Don't admire yourself in your side view mirror or do anything else narcissistic.

When you first meet the initial interviewer, he or she wants to see someone who appears pleasant, friendly, and prepared. This initial three-second subjective assessment sets the tone for your interview, so look him straight in the eye and smile. He'll subconsciously credit you with being bright, enthusiastic, and businesslike. Congratulations—you've just squashed his biggest fear *and* put him at ease because you're perceived to be just like him!

Employers want "can do," "will do" employees. An Employer considers the following factors:

- ◆ Character.
- ◆ Loyalty.
- ◆ Initiative.
- ◆ Personality.
- ◆ Communication skills.
- ◆ Experience.

Throughout the meeting, the interviewer will be determining whether or not you have these traits, largely according to how he perceives your body language. Here are some techniques you can use that will help you appear to be the kind of person you really are, and that the interviewer is looking for, even when you're nervous:

- ◆ Sit up straight, but not rigidly.
- ◆ Acknowledge key points made by your host with a nod, smile, or other appropriate gesture.
- ◆ Maintain good eye contact, without staring down the interviewer.
- ◆ Don't cross your arms in front of you (you'll appear to be unfriendly); keep your hands folded in your lap or balanced on your knees. Avoid touching your face.
- ◆ Avoid nervous habits, such as tapping your fingers, jiggling your feet, rubbing your nose, pulling your ears, twisting your hair, or picking your nose (you have no idea how many people do this without noticing).

Basic Interviewing Strategy—The Old One-Two Punch

There are only two ways to answer interview questions: with a short answer or a long answer. They follow each other in sequence as a good one-two punch. For an open-ended question, I always suggest you say, "Let me give you the short version. If you need to explore some aspect more fully, I'd be happy to go into greater depth, and give you the long version."

Why is this important? Short answers of less than 30 seconds are generally insufficient, but answers of more than three minutes are too long. In the first instance, you'll come across as light, lacking knowledge, depth, and insight. Long answers right off the bat, though, could brand you as being too technical or boring. You need to strike a balance, and this one-two punch is the best way.

The reason you should respond this way is because it's often difficult to know what type of answer each question will need. A question such as, "What was your most difficult assignment or biggest accomplishment?" might take anywhere from 30 seconds to 30 minutes to answer, depending on the detail.

Therefore, you must always remember that the interviewer is the one who asked the question. So you should tailor your answer to what he or she needs to know, without a lot of extraneous rambling or superfluous explanation. Why waste time and create a negative impression by giving a sermon when a short prayer would do just fine?

Let's suppose you are interviewing for a sales management position and the interviewer asks you, "What sort of sales experience have you had in the past?"

Well, that's exactly the sort of question that can get you into trouble if you don't use this method, because you won't know where to start.

You might answer the question in this way: "I've held sales positions with two different software companies over a nine-year period. What aspect of my background would you like to concentrate on? Where would you like me to start?" Develop a neat concise package.

By using this method, you subtly telegraph to the interviewer that your thoughts are well organized and that you want to understand the intent of the question. After you get the green light, you can spend your interviewing time discussing, in detail, the things that are important, not whatever happens to pop into your mind.

Behavioral Interviews

> *Describe the most difficult decision you have ever made, and explain your approach to the decision-making process.*
> *What decision did you have to make?*
> *What options were available to you?*
> *What factors did you take into consideration?*
> *What did you decide?*
> *Why do you think that you exhibited sound judgment in this instance?*

If your interview begins with a question such as these, you could be in for a behavioral interview. I hope you're ready. Get ready for a grilling! Based on the premise that the best way to predict future behavior is to determine past behavior, this style of interviewing is gaining wide acceptance, especially in the technology industry.

Today, more than ever, every hiring decision is critical. Behavioral interviewing is designed to minimize personal impressions that can affect the hiring decision, by focusing on the candidate's actions and behaviors, rather than on subjective impressions that can sometimes be misleading. It facilitates more accurate hiring decisions.

If you have experience with traditional interviewing techniques, you will find the behavioral interview quite different in several ways:

◆ The interview will be a more structured process that will concentrate on areas that are important to the interviewer, rather than allowing you to concentrate on areas that you think are important. (You can figure out what's important to him or her by carefully reading the ad and the job description.)

- Instead of asking how you *would* behave in a particular situation, the interviewer will ask you to describe how you *did* behave. (You need to be able to flip back to that instance in your mind quickly.)
- The interviewer will ask you to provide details. This is not the time to theorize. (That's why you wrote out all those accomplishments in Chapter 1.)

How to Prepare for a Behavioral Interview

Practice, practice, practice. A behavioral interviewer has been trained to collect and evaluate information objectively and works from a profile of desired behaviors that are needed for success on a job. Because the behaviors a candidate has demonstrated in previous similar positions are likely to be repeated, you will be asked to share situations in which you have exhibited these behaviors. These questions and your answers will be directly compared to a reference check form, so don't lie or "wing it," because you'll be caught.

Preparing for a behavioral interview is much like writing out your greatest accomplishments, only you need to focus on the duties to be performed in the job instead. You can pull out those duties by reading the ad carefully. Based on these you need to recall recent situations that show favorable behaviors or actions, especially those involving leadership, teamwork, and initiative. Prepare short descriptions of each situation. Be ready to give details when asked. (See Appendix I for sample questions that were graciously culled [with permission] from the database at the Software Human Resource Council [*www.shrc.ca*].)

Traditional Interviews

Most traditional interviews progress through three distinct stages:

1. **A warm-up general discussion.** Mostly likely this will be with someone from the HR department. You will be questioned about specific jobs you've held in the past that are of interest to him or her—not necessarily you. The more time he or she spends on any previous job, the more relevant it is to the position you're being interviewed for. Don't be in a hurry to move onto something you think is more interesting. You're not running the show.

    Be prepared to drill down on each job and show how your experience fits with what he or she wants. Bring your success stories forward. Ply the interviewer with analogies. Ask, "Is there anything else you'd like to know about this?" or "Does that answer your questions?" to make sure he or she got the details they need. Make a mental note to emphasize this in your thank you note.

2. **A more in-depth technical discussion.** This is especially prevalent in engineering jobs, but it also applies for finance and sales. You need to demonstrate your up-to-date industry knowledge around the particular technology he or she is looking for. If he or she is hiring for a "Blue Tooth" project, your knowledge of MS-DOS isn't particularly relevant.

    Following the chat board discussion on Blue Tooth standards is relevant though. Salespeople need to be prepared to drill down to the tactical level to explain the who, what, where, when, why, and how of your major accomplishments. Techies in most cases will be interviewed by a fellow techie who will be trying hard to demonstrate his or her technical prowess as you are interviewed. Bone-up on the latest "buzz words" before you go.

3. **A pleasant closing discussion.** Finally, after he or she has interrogated you for an hour or more, he or she will ask you if you have any questions. Usually he's just trying to be polite and doesn't expect any. This is the moment of truth where you can outdistance your competition and/or make up for a poor interview.

Asking the Right Questions

Your personal question period separates the men from the boys, so to speak. This is your only real chance to put serious distance between you and anyone else who may being considered. As Thomas A. Freese points out in his book, *Secrets of Question-Based Selling,* "Asking questions puts you in a desirable position with respect to your sales conversations. It puts you in control....You can control the pace and depth, depending on what questions you ask and how you ask them. If your prospects are asking all the questions, then they are controlling the sales process."

Ask the right questions and you will lead the employer to the inevitable conclusion that you are the right choice. Think about this for a minute: When someone asks us deep thoughtful questions, we think they're smart and important. This strategy can create demand for you, and it will absolve you from all your previous blunders. If your interview didn't go very well or was only so-so, here's where you make up ground and overtake your competition.

Your question period is the most important event, because your copious research will become apparent. You can learn the company's weaknesses and map them to your accomplishments to show you can provide the solution. Employers often stop interviewing and start nodding their head up and down in complete agreement.

When you start to see that silent side-to-side head movement, don't panic! This doesn't mean no; it means, "I can't believe this guy understands my problems." You might get an offer on the spot. Resist the temptation to take it. Ask all your questions first and take notes. An occasional "that's interesting" followed by note-making will make the interviewer squirm. The interviewer will think, "I should have been selling. He's been assessing ME the whole time! I can't let this one get away!" He or she will figure you're a "sleeper" and panic. Watch the interviewer start to sell you—and sell you hard!

Once you are at this point the simplest way to close the interview is to finish all your questions and "gently" drill down on the answers. When you're just about ready to go, ask one last question: "Is there anything else you think I should know about the company or this position?" This signals that you've finished.

If the interviewer likes what he or she has seen, he or she will say something else—it doesn't really matter what—to take back control of the interview. Tell the interviewer you're interested in meeting some of the key people you'll be working with if he or she thinks there's a good fit. **Stop talking—not another word. That long, silent pause has to be filled by the interviewer, not you!**

As you are leaving the interview, thank the assistant. If he or she hasn't already told the boss how you called to confirm the interview, this will be pointed out after the interviewer comments on the grilling you just gave them.

Lastly, if there are other candidates waiting just smile at them as you walk out and watch them squirm. You probably ran overtime into their interview schedule. You've just set the bar so high the rest of the people interviewed will be held up against you as the benchmark. As I like to say, "When you're interviewing, you are either part of the steamroller or part of the pavement." Make sure they are the roadkill, not you!

Guidelines for Asking Questions Correctly

Write down questions that are important to the business, market, and success. You have to show you understand its Value Table. Bring a notebook of some sort that looks professional to ask your questions from. And jot notes. It's rude and illogical to ask deep probing questions and not take notes. You don't want an instant disconnect at the front end of this session. You could be considered a phony, whether you are or not.

Prepare five good questions, understanding that you may not have time to ask them all. Ask questions concerning the job, the company, and the industry. Your questions should indicate your interest in these subjects and that you have read and thought about them. For example, you might start, "I read that _____. I wonder if that factor is going to have an impact on your business."

Don't ask questions that raise warning flags. This is a question period, not an interrogation. You want to impress the interviewer with your insight, not start a debate on development methodology or the company's accounting policies. Don't ask questions about compensation (pay, vacations, perks of any kind). You will be viewed as being more interested in paychecks or time off than the actual job.

Clarify. It's okay to ask a question to clarify something the interviewer said. Just make sure you are listening. Asking someone to clarify a specific point makes sense. Asking someone to reexplain an entire subject gives the impression that you have problems listening.

10 Questions That Will Force Any Technology Employer to Sit Up and Take Notice

1. Can you explain to me how your business philosophy has changed/evolved over the last five years? How does this compare to your competitor X?

    > *The answer tells you what the company values most. It will tell you whether the company is product- or market-driven and where its weaknesses lie. If the company hasn't changed in five years, it is either a run-away success or "blind." Either way, your opportunity to make a real impact could be minimal.*

2. How does the company deal with the inherent conflicts between quality and the timely delivery of new products?

    > *The answer tells you how realistic the company is and whether its alpha- and beta-testing processes are well run. It also gives you insight into how the different departments operate. Anything less than total cooperation between kingdoms is a recipe for disaster. Everyone works on "Internet time" now.*

3. What industries or outside influences affect the company's growth? How have they affected X company?

> *The answer tells you how the company minimizes the downside and maximizes the upside. No one controls a market (forever), and many external influences can impact a company's success. A smart company will be able to recognize these external influences and leverage them. If it doesn't know what its competitor is doing you've got a problem.*

4. What are five major short- and long-range goals and objectives? How is success/progress being measured? How is the company doing against these metrics?

> *The answer defines the company's vision and underscores its grasp of the market. Do you think the company is right? If not, why not? Do you think it is geared for the long run and has the financial muscle to accomplish its goals? Is the product/service really spectacular or just average? How can your experience help the company achieve its goals?*

5. From an overall effectiveness standpoint, what might improve or enhance the company's competitiveness?

> *The answer tells you whether the company knows its weaknesses; if it doesn't, it is headed for trouble. How do your strengths and/or interests play into this? Can you really contribute?*

6. What are two or three characteristics that your company feels are unique?

> *The answer tells you what kind of people the company attracts. This is a trick question: If the company hires no one but high-energy, motivated, positive people, it could be headed for trouble. A company needs people whose opinions clash, especially on major issues such as new product development. If not, companies have a tendency to fall in love with their own technology.*

7. What are the three main functional tasks of this position?

> *The answer tells you about the heart of the responsibility. It also throws light on whether or not you'll have the authority to actually complete your responsibilities.*

8. What is the turnover rate in the engineering and sales departments?

> *This answer tells you how well the company hires. Explore this carefully. High turnover is neither good nor bad. In fact, low turnover could be more damaging than the latter. Compare this to the company's growth curve and new-product-development cycle. Be suspicious of a sales department with low turnover. There should always be a 10- to 15-percent movement here, unless the product is in super demand or people are beating down the door to sell for them. Extremely high turnover (40 percent) in either the sales or engineering department could indicate that the company doesn't hire well to start*

with (which isn't uncommon) or that the company lacks focus and isn't bringing new products to market in a timely manner. Will you always be understaffed?

9. In what stage of the buying cycle is your product?

The answer tells you whether it's an early adaptor product/service or if it is seeing its sunset. Your ability to sell (if that's what you do) in a missionary role—to convert the non-believers—could be critical to the success of this company. If you're more of a farmer, you may feel uncomfortable in this role. If you like to build new products and play with state-of-the-art tools, a sunset firm will be dull and you'll soon be looking again.

10. What is your debt-to-equity ratio?

Odd question, you might think, but in reality any company with a debt-to-equity ratio more than 3:1 could experience difficulty raising cash to launch a new product, no matter how hot the market. The dot-com bust proved that!

The Psychology of Money

Most people hate talking about money. Not me. It's the most natural thing in the world. Having said that, money is about the only issue discussed during interviews in veiled phrases. Too many people shoot themselves in the foot at the last minute when it comes to this. Unfortunately, you're finished as a candidate if you can't answer the question when the employer finally asks, "How much do you want?"

If you pause, stumble, fumble, or mumble, the interviewer will assume you're playing him or her for more money, for more time to assess another offer, for whatever. It's just like telling someone "I love you" for the first time. You just sit/stand there and wait and hope and wait to see if he or she loves you, too. When he or she delays, or answers cautiously, you know how humiliated and angry you feel (I am not projecting here this is really what happens). Employers tend to become offended when this happens. It's in your best interest to be able to answer the question whenever you get asked. There are only two things you can say:

"I want X."

Or the more appropriate response:

"I like this company and I like the opportunity, but it's premature to discuss potential compensation until we've mutually agreed that this is a good fit for both of us. Wouldn't you agree?"

Practice this line until it rolls off your tongue so smoothly you could be nominated for an award. It must sound natural.

This answer brands you as savvy, bright, and capable. Now if he or she says right after that, "We want a range to make sure we're in the ballpark" or something to that effect, then you can probably figure he or she is interested—but you never know, so again, just say, "I'm certain you'll be fair."

The interviewer won't ask again. If he really is interested, you've put the responsibility on him and the interviewer won't want you to get away, so he may start tossing out hypothetical numbers. Put your hand up and say, "Jim, there are a couple of more people I'd like to meet first but I'm very interested. The dollars will work themselves out. Please, may I talk to Sally now?"

Voilà! Do your last interview and expect an offer in writing

If this sounds unusual, it isn't. You just won't get this advice from a career counselor. They will tell you to state a wide range and hope you land somewhere in it and that the company will offer the top end. That doesn't happen in real life!

The Follow-Up

A follow-up letter thanking the interviewer for the meeting and expressing an interest must be written the same day. My mentor, Bob Henault, years ago taught me to send a quick three-paragraph note expressing your sincere interest in the company with three bulleted sentences covering, "You need-I've got"as well as a simple call to action, such as, "What are your thoughts on a second meeting?"

You will be remembered as the "guy who called ahead, too," and it will separate you from the pack.

Knocking out the competition after the interview is over is tough. We recruiters can sell our candidates, but a candidate selling himself is much more difficult and nearly impossible. These days, just sending a thank you note and waiting for something to happen doesn't get the job done. Calling and asking questions on the stage of the decision process lets the employer know the candidate is interested but does it make him or her stand out from all the others? No, not really. Shari Miller, principal of The Elmhurst Group (*www.elmhurstgroup.com*) instructs her candidates to draft a proposal on how they would handle some of the position's requirements. Also, in their follow-up note she suggests they inquire as to the next steps or request to get together to discuss their ideas on a subject they know is of interest to the employer and relevant to that job. With unemployment so high, beating the competition is not as easy as it once was.

Shari graciously agreed to let me to include the following two examples. The idea is to convince the employer that you can solve its problem by demonstrating how you solved those of a previous employer. You lay out the situation, action plan, and results and then ask for another more detailed meeting to get at the information you need to do one for this employer. This type of follow-up and due diligence will quickly set you apart from the other candidates and assure you of another meeting (and probably an offer as well).

Example 1: Successful Sale of an Internet Company During the Dot-com Bust

Situation

During the frenzy of the venture firm–backed dot-com era, the company had raised $23 million in venture capital based on a unique proposition but weak on underlying business principles and execution. The final round of financing was completed in December 1999, just four months before the dot-com bubble burst.

Upon my arrival in June 2000, only $8.3 million in cash remained of the $17 million final round. In addition, there was $300,000 in secured debt owed to a financial institution and a $2.5 million convertible subordinated note due to a financing company.

As a result of poor execution and management, the board terminated the company founders and installed a new CEO just 10 months prior to my arrival. With the new CEO the company achieved traction in driving subscribers; however, the business model sponsored a free service deriving revenue instead from the sale of advertising and, as a result, was very limited. Based on the conditions of the company and perhaps more on the current marketplace, the board concluded that in order to achieve any return it would be wise to sell the company. Therefore, the CEO was directed to reduce overhead and put together a plan to sell the company. My services were engaged primarily to lead the effort.

Action Plan

- Reviewed existing financial statements, budgets, promissory notes, and subordinated debt instruments, private equity financings, contracts, and all agreements to assess the underlying principles and to determine the availability of options to raise revenue and reduce costs.

- Collaborated with the CEO to develop an action plan to drive membership, increase revenue, implement a new budget, and preserve the existing employment base in order to maximize the attractiveness of the company to a suitor.

- Compiled a comprehensive list of potential companies and prepared presentation materials for meetings with prospective acquirers.

- Renegotiated or terminated existing agreements, brokered business development deals, and met with business partners to increase advertising revenues and reduce costs.

- Maintained open communication with employees to maintain focus on job performance and keep them informed on the progress of financing negotiations.

Although the board had originally requested a reduction in force to cut costs, we convinced them it would be counterproductive to achieving maximum value for the company.

Results

After numerous meetings with various companies, management identified, negotiated, and successfully closed the sale of the company to a European competitor listed on the German Stock Exchange. The company was successfully financed and saved the investors from a certain write-off. In the new era of the dotcom fallout, the sale was a win-win for all parties concerned. The investment group received publicly traded liquid stock with an opportunity to recoup the original investment and achieve a return. Creditors, vendors, and employees won because they had potential with a much stronger and more viable company. Subscribers won because they now had a much broader subscription base with enhanced services to draw upon, and, lastly, the acquirer doubled the size of its company overnight with the acquisition of a new subscription base plus enhanced technology.

Example 2: Successful Completion of a Challenging Initial Public Offering

Situation

An 25-year-old retailer had grown its operations to 50 stores in four states, generating $32 million in revenue. Although the company was profitable, it was experiencing numerous problems, including weak or nonexistent reporting processes, antiquated and cumbersome operating and information technology systems, inadequate and expensive credit facility, and thin and inexperienced management. The company's founder aspired to take the company public in order to raise capital for expansion and personal liquidation. Further, he was steadfast about selling only one-third of the equity, thus retaining two-thirds ownership. Due to his naïveté, investment issues were handled poorly. Only one investment banker had been chosen to underwrite the transaction, and no other sell-side analysts had been enlisted to write coverage on the company. The banker insisted on a very short window between the filing of the initial prospectus to the "road show" itself. This left little time for conducting due diligence, writing the S-1, and completing audits for two financial statement reporting periods while maintaining the ongoing operations with a lean and inexperienced financial staff.

Action Plan

◆ Assessed and reorganized finance staff, including replacement of incompetent employees with new middle-management hires to address ongoing operations and realignment of responsibilities. Reinforced marketing staff with additional experienced personnel and enlisted their support in selecting and implementing new information systems.

◆ Cultivated and implemented a new, more responsive commercial banking relationship to replace an ineffective and costly predecessor.

◆ Through personal relationships, contacted other investment banking and securities management firms in order to generate interest in the company's offering and expand underwriting and analyst coverage.

◆ Laser-focused on completing audits and filing S-1 within established time frame.

◆ In conjunction with outside consultants, developed a plan to replace outmoded information systems as soon as the initial public offering was completed.

◆ Engaged an executive coach to assist founder with his limited public-speaking capabilities.

◆ Personally took the lead role for the majority of the "road show" presentations.

Results

The company successfully completed and executed its initial public offering within the original time frame specified by the investment banker, raising more than $16 million. This was quite an accomplishment, given that capital markets had suffered another "black" October, with the Dow Jones Industrial Average dropping more than 554 points or 7 percent in a single trading day. Diligent investor relations and conference presentations resulted in engaging three other analysts in writing coverage and providing recommendations for the company. The company successfully negotiated a new $20 million credit facility with a major financial institution that provided superior borrowing terms, reduced covenant restrictions, and improved retail service for the stores. Post-IPO, the company implemented an enterprise-wide fully integrated information system within a six-month time frame that vastly improved reporting processes, enhanced customer service and information gathering at point-of-sale, and overall inventory management, contributing to a 25-percent improvement in inventory shrinkage. This improvement created an opportunity to enhance television advertising and increase spending to drive sales. Lastly, the company experienced a 13-percent pre-tax savings in interest costs in the initial year following the IPO.

References: It's Offer Time—Almost!

So you had a great interview and the employer wants to extend an offer—after he or she checks your references. Congratulations? Not yet: Here's where most job offers are lost. That's right, most job seekers lose their job offers at this stage because they're lukewarm or neutral. And it's your fault! But you can also fix it.

So how do you use references to your advantage? By understanding from the employer's view why the employer needs to do them and what the employer is looking for. With a little forethought, you can help your references help you, but first let's talk about why references are being checked more often these days.

In the Steven Spielberg film *Catch Me if You Can*, Leonardo DiCaprio portrays Frank W. Abagnale, a master impostor and forger. Abagnale worked all over the world under a number of false identities—a doctor, a lawyer, a college professor, and as a copilot for a major airline company—all before reaching his 21st birthday. It's a very funny and entertaining movie that begs the question, "How did this guy ever get hired?"

It wasn't that hard. Obviously no one checked his references. Con artists, masters of deception, and people right in your hometown fool the hiring system every day, robbing the rest of us and putting people's lives at risk. Obviously no one checked Frank's references. Forty years ago a job seeker could get away with that because of technology limitations we don't have today. Information can be had almost instantly, so there is no excuse for not checking references.

The following statistics underscore a universal problem: the incredibly expensive price and high stakes consequences of letting the wrong person into your organization. According to the American Management Association and the U.S. Department of Justice, employee theft and dishonesty:

- Cost U.S. businesses between $60 and $120 billion per year (not including the billions spent on protecting against theft guards, security systems, and so on).
- Cost $36 billion annually for workplace violence, and the "average" employee embezzles $125,000 over the course of his career! (And that's the "average" person; imagine what an organized guy like Frank could do.)

So, that's why references are critical to good hires and companies have come to appreciate this even more in the post-boom economy.

Over the past 20 years I've done at least 10,000 interviews and, even for a professional recruiter such as myself, it takes a lot of work and preparation to see past the veneer of most job seekers. It's very difficult to get candidates to be themselves. Most candidates come dressed for success and prepared to dazzle the interviewer, knowing or at least hoping that their references will never be called.

During the 1990s the high-tech sector in particular turned the entire hiring process into nothing more than one-sided sales pitches where interviews and reference checks were a formality and rarely done for fear of losing a hot candidate. "Time to hire" was the mantra of the day and for a while it seemed every candidate had an agent or personal spin-doctor. Instead of checking references, employers relied on probationary periods to weed out the duds. Back then candidates just had to remain civil for a few months. Then the market crashed and the rules changed.

After the Crash

First, employers cannot afford to wait and see if a candidate works out. They are strapped for cash and time so they will not waste either with a so-so hire. Second, the recent proliferation of resume factories, coupled with higher levels of unemployment, have led to the situation that some HR practitioners call "credential-creep." Credential-creep is the fine art of exaggerating. Sometimes it even means out-and-out lying on a resume on the off chance that an employer won't or can't check a reference to verify all the details. The fact is that people lie on paper when they're looking for a job. In a tight economy the slightest edge is often what makes the difference between securing an interview or being filed in the circular filing cabinet.

Credential-creep can range from a candidate claiming he or she had greater responsibilities than he actually did to claiming he worked for a company when in fact he did not.

To illustrate that fact, let me tell you about the experience of one of my clients: Brian, one of my clients in the United Kingdom, called me shortly before Christmas. He was preparing to hire a fabulous candidate he'd met at a party and wanted my input on the offer. On paper, the fellow was a dream come true. A sales vice president with 20 years' experience from three of the world's largest systems integrators. Pedigrees such as that are rare. The nice chap had even provided written references from the presidents of these firms. They were sterling—a little too good, actually.

I insisted on checking the references myself and, because I had worked with him for more than 10 years and I had hired more than 20 of his key staffers, he agreed. It turned out that "Reginald," or "Reggie," as he referred to himself, had a deep and rich fantasy life. He had made it all up. Everything. None of the presidents had even heard of him. When confronted with the lie more lies were told—something about being undercover with MI-5. Needless to say we didn't hire him. If we had it would have cost us about $100k in salary, termination costs, and legal fees for those three months—not to say the least about the lost time and aggravation.

Now this may be an extreme case but people do fudge, embellish, augment, and incorrectly describe their credentials. Everyone who has hired more than once knows that, as an isolated activity, interviews are unreliable in predicting success on the job.

Why Interviewing by Itself Is Unreliable

There are so many books and success coaches nowadays to help coach or train a candidate in how to perform during an interview that most job seekers are more skilled at interviewing than the interviewers themselves. Interviewing during the go-go 90s was as much a spectator sport as a tool in the hiring process, and the dice were loaded in favor of the job seeker.

For most employers who do occasional interviews the process is a farce. It reminds me of a gentlemen's tennis match where everyone is dressed in white and is careful to remain immaculate. Questions are asked; questions are answered. Back and forth, forth and back, with very little in-depth discussion. The mood is light and congenial—nothing like real life.

Referencing is often the last, and indeed the only, way to separate out who's really got the skills and who's just a fast talker. Frequently interviewers extend offers based on their first impression, or "gut feel," or "chemistry," with little regard for the hard evidence that proves which candidate is the right one for the job.

So if this isn't enough to convince you to be honest because employers will reference check every candidate, then let's talk about an employer's legal obligations.

The Legality of Reference Checking

Can an employer be held liable if they don't check references? You bet. As a matter of fact, as if it isn't tough enough just to make a profit these days, many companies have been held liable for crimes committed by their staff members that have ranged from theft to murder. This new legislation called Negligent Hiring, varies from state to state and province to province in Canada.

The law means a company can be held liable for failing to conduct an adequate pre-employment investigation into a job seeker's background. If an employee has a history of misconduct indicating a propensity for violent behavior or other misconduct, which an employer could have discovered through a reference check or other background investigation, the employer could be held liable for any and all resulting injuries. Failing to adequately investigate before hiring can expose the employer to liability for actual injuries, pain, suffering, and even punitive damages.

Of course, you are not the type of person who would ever do anything wrong, but the employer does not know that and cannot risk not checking you out. "Okay, Dave," you say, "I'll just make sure my references say nice things," or "It doesn't matter because my employer has a no-reference policy so if they call they will not discover anything negative." Let me tell you, that won't work because most states grant immunity for good-faith references.

References can be held liable if they give a false reference or bury the truth. In another widely publicized case a lawsuit was brought against Allstate Insurance Co. This was a "negligent referral" case. The suit was settled before going to trial, but a Florida judge ruled that Allstate could be sued for punitive damages for concealing the violent nature of a former employee who killed coworkers at Fireman's Fund Insurance Co.

In this case, the wrongdoing allegedly occurred when Allstate wrote a recommendation letter saying the employee was let go as part of a corporate restructuring. In truth, he had been fired for toting a gun at work. In January 1993 this man shot five Fireman Fund coworkers in the company's cafeteria. He killed three of them before fatally shooting himself. One of the survivors and the families of the victims filed the suit against Allstate. Fireman's said it relied on the letter from Allstate when it hired him.

Yes, this is an extreme case, but it is a reality that keeps employers up at night. In the United States, 35 states have passed laws that protect employers by granting them immunity from civil liability for truthful, good-faith references. Although the laws vary by state, the statutes specify that an employer will be presumed to be acting in good faith unless the current or former employee can prove that the reference provided was knowingly false, deliberately misleading, malicious, or in violation of civil rights laws.

Reference checks are now pivotal to the hiring process. I have pulled more than my share of offers because I discovered inconsistencies in a job seeker's resume or interview, and I am far from being alone on this issue. Now that you understand why they are important you need to know what specifically an interviewer is going to ask. But let's first talk about who they are going to call.

Who Is the Right Reference?

Employers know full well that most job seekers stack the deck in their favor when asked for references. Unless they specify with whom they want to speak, they know your list will be filled with people ready to sing your praises. If you really, really, *really* wanted a particular job, wouldn't you do the exact same thing? Okay, maybe *you* wouldn't, but most people would. Most people don't want to risk having someone tell the truth.

So who's a good reference? It's not your priest, best friend, or drinking buddy. It's probably not your wife either. Employers need to talk to those in a position to judge your competency to do the job you are being hired for.

Who is a good reference?

The best referees are those that have recent, firsthand experience of your performance on the job—most often line supervisors and peers. Don't use personal references. It's not just that their opinions don't count (and they don't); it's that they can't possibly add any insight about your work habits on the job. Your "buddies" don't know what you are like as a fellow worker. *Character references have their place, but it's not here.*

Employers have two basic objectives. First, they have to make sure Mr. "Star Candidate" really can do the job. References can help to substantiate or nullify facts and impressions gathered from your interview. Second, they need to make sure you aren't a gun-toting axe murderer, to protect their other employees and their customers too.

After talking with your references, this is what they will be gathering information and assessing:

- ◆ Significant accomplishments.
- ◆ Problem-solving skills.
- ◆ Leadership styles.
- ◆ Strengths and weaknesses.
- ◆ The depth of others' feelings, positive or negative, about you.

- Management guidance or further professional development required.
- Management and personal style.
- Relationships (internal and external).
- Depth of technical and professional skills.
- Career progression and career interests.
- Reasons for changing jobs.

Here are some of the basic questions they are likely to ask:

- What were the candidate's strengths on the job?
- Were there areas in which he or she should improve?
- Was he or she dependable?
- Was he or she a team player?
- How would you compare his or her work with others who held the same job?

If you pass these basic questions they will ask about specific projects you discussed during your interview and compare the reference's answers with yours for consistency. For example:

- What was his or her biggest accomplishment at ABC Corp. in your opinion?
- What do you anticipate I will find to be this person's real strengths, and what areas would benefit from constructive coaching or mentoring?
- This position interacts with XXX types of coworkers or customers in YYY types of situations. How does that compare to what he or she did for you, and how well do you think he or she will handle these interactions for me?

Here's How You Can Help Yourself

Obviously there is a lot of work involved in doing a proper reference check. At my firm, Perry-Martel International Inc., we ask each reference 68 questions during a standard reference check (I've made one bad hire in 15 years), and I am amazed at the lack of foresight of most job seekers who let their references be caught off guard. Here's what you need to do when an employer says the golden words, *reference check*:

1. Call your references and tell them who will be calling.
2. Get into their schedules and confirm times.
3. Get after-hours phone numbers and cell phone numbers if you can.
4. Talk to your references and tell them why you are excited about the new opportunity. Let them sense your excitement, because they will transmit that to the person checking the reference.
5. Tell them how your background fits the employer's needs.
6. Remind them of your accomplishments that are relevant to the position.
7. Lastly, send each of them a copy of your current resume and a copy of the job description for the new opportunity, and highlight where the fit is.

Why take chances? Now that you understand a little better why and how they will be checking your references, you can position yourself to get the reference you deserve. A little preplanning can save your offer and sometimes even get you a better offer. Most people will not go to this much effort, but what would happen if you were in a tie in the employers mind with another candidate and you did this? An ounce of prevention, as they say.

Chapter 6

Finding and Using
a Recruiter

If you want to move ahead in your career, you just can't ignore the new world of recruiting, which has been driven by a strong demand for technology people worldwide. Boosted by a fickle job market that has employers seeking outside assistance to find workers, North American staffing industry revenues reached $140 billion in 2000. The industry may be weak at the moment, but it will come back quickly when the market picks up.

In 2000, there were approximately 125,000 professional third-party recruiters in the United States and several thousand in Canada. Whatever euphemism you use to describe them (recruiter, headhunter, executive search professional), learning how to work with them is a skill you need to master.

There are two types of firms: retained and contingency. Retained firms generally focus on senior executive searches and are paid by their client company for working on the file regardless of whether or not the search is successful. A portion of the fee is normally paid up front, with the balance due over a predetermined time. Contingency firms are paid if, and only if, a client company hires one of their candidates. Generally both firms are paid a percentage of the first year's income. In other words a $100,000 a year salary will fetch the recruiter $30,000 at a 30-percent fee rate. Fees can range anywhere from 5 to 35 percent or more. Retained firms concentrate on positions paying in excess of $100K and contingency firms on less than $100K.

The major difference you need to be aware of is this: Contingency firms may decide to proactively market you to their clients or to a specific geographic market. They will assume all the costs and all the risk in doing so retained firms generally will not do this. So what's this mean to you? Simply put, if you decide you want to have a recruiter market you, you better be:

- Outstanding in your area.
- Able to prove it.
- Cooperative.

General Rules

1. Find a headhunter *before* you need a new job. This is the most obvious and most important rule. You're probably wasting everybody's time talking to a headhunter unless you're at least thinking about making a job change, but the absolute worst time to make first contact is when you need a new job right away! The purposes of a first contact are to let the headhunter know that you exist, give the headhunter some idea of your talents, and let him or her know you're sort of looking around but you're in no big hurry. You need to have a six-month runway in front of you to use a headhunter strategically.

2. If you're employed, don't be too quick to send your resume to an unknown headhunter. Ask around. Ask your friends for referrals. Ask your former HR department. When you finally find a good one, your goal is for that headhunter to keep you in mind for career-building positions that he or she has been engaged to fill. When you're solicited "cold," inquire as to whether the headhunter has an "exclusive." If he or she does, he should have no problem telling you about the company. They often are just fishing, trying to find a skilled candidate he can package and market. Although there's nothing unethical or wrong about that, you really need to know exactly what he will be doing with your resume. Having your resume "shotgun blasted" to every one of your current employer's competitors is a career-limiting move. I once worked with a guy (for a very short time) who had 400 employers programmed into his fax machine, and the minute you left his office your resume went out, with or without your approval, with predictable results.

3. Be prepared to tell the recruiter why you deserve to be designated as an "MPA," or Most Placeable Applicant. To be an MPA you need to have a well-honed and rare skill set. Being a C++ programmer isn't rare; being a mainframe Java programmer with five years of real-time experience might be. Most financial people are "vanilla" unless you have U.S. GAAP experience and have done an IPO on NASDAQ. I am sure you get the idea. So what do you do if you aren't as rare as the guy in the cubicle right next to you? Have a great personality and cooperate! If you actually look as though you can follow instructions and will work with the recruiter, he or she may invest a few hours or days in marketing you. To hone your interview skills you might also consider enhancing your presentation skills by attending six months of Toastmasters (*www.toastmasters.com*) luncheons. It's an inexpensive program with huge dividends for you personally.

4. You must find a headhunter who specializes in people with your talents. Great recruiters specialize by either geography, technology, industry, or some combination. Find a headhunter who specializes in your technical field and geographical preference. Don't bother sending your resume to all 125,000; the 'delete' key files irrelevant resumes instantly. Go to *Google.com* and put in your city, industry and the word *headhunter*. Voila—more headhunters than you can shake a stick at. Now go and read their sites carefully.

Find those individuals who specialize in your skills. Log onto the Net and check out Kennedy Publications for its list of North America's top headhunters (*www.kennedyinfo.com/er/erindex.html*). Kennedy is *the* authority.

5. Provide the headhunter with your salary expectations. Most people find it hard to discuss compensation and that's fine. Not me. It's the most natural thing in the world for me. It's usually the third or fourth question I ask. Judging by the looks I sometimes get, I am an exception. People like to keep this information private. There are, however, at least three people in this world who need to know your current salary: the IRS, your spouse and your headhunter. Headhunters do because we need to *exclude* you from all jobs *under* your salary level. If you play cat-and-mouse with us you'll end up in the circular filing cabinet and brand yourself as a primadonna! The entire firm will take a hands-off approach to you. Interestingly enough, most headhunters don't use your salary to peg your seniority. Years ago I met a great marketing director who was making $38K. She was grossly underpaid and underappreciated. She went on an interview for me and got an offer for $92K. She was floored, and it took me two days to convince her that she was worth that amount. (By the way, my client's upper end was $125K, so he didn't overpay for her skills.) More importantly, if he had known what she was making at the time he wouldn't have made an offer because he would have concluded she was too "light" for the position. She's done a fabulous job and has been promoted to vice president.

6. Don't put important facts in your cover letter and then not put them in the resume. Cover letters get lost and tossed in the shuffle. They only keep the resume. If you are sending an unsolicited resume get to the point fast. You have somewhere between five and six seconds to convince a headhunter to read a resume he or she didn't ask for. Don't bother saying you have heard from a friend what a great recruiter he or she is unless you can name the friend. You will appear phony.

7. Make it easy for a headhunter to get and read your resume. Don't assume that everyone has every possible technology in place to receive and read your resume. We sometimes get, "Use your Web browser to log onto my personal Website and download a copy." Forget it. Send it as an attachment in Word or in the body of your e-mail message. Do not send an Adobe PDF file unless requested to do so. This is not as common a technology with recruiters as you probably think; most still struggle with Word. No one uses WordPerfect. Only use industry-standard technology unless told otherwise.

8. Headhunters are usually *not* good vehicles to help you change careers. Employers use us to help them find people with a particularly well-honed skill set and several years of practical experience in a particular field. If you're an unrecognized "superstar," headhunters are usually quite good at helping you get recognized and accelerate the advancement of your career. We are not very effective at helping you change careers. Read *What Color is Your Parachute* instead. It is truly the best book around for career changers. Headhunters won't care about your transferable skills; they'll have six other candidates already who don't need morphing.

9. If you know the game ahead of time, your odds of winning are greatly increased. Headhunters recruit; career counselors counsel! Headhunters such as myself are constantly inundated with requests for free career counseling, free resume writing advice, free practice interview sessions, free job leads, and so on. Although most of us are more than

qualified in all of these areas, our "real" job is to hunt heads for our client companies. Don't expect you're doing someone a favor by letting us "have a look at your resume early" or "getting the opportunity to market you." That's not how we feed our families.

10. Headhunter is not a pejorative term ("flesh peddler" is). You won't, however, find the term on most recruiter's business cards, so when in doubt call them executive search professionals—ESPs for short. Most good ones have ESP anyway—it goes with the territory. Working with a professional headhunter can be a very rewarding experience for you and your career. We tend to be connected to the people you want to know. Any recruiter who has been in the business for more than two years should be considered a battle-hardened professional.

Just remember this: If you want to be found, you need to be visible. We will find you even if you're hidden out of site but then again we might find someone better with less effort, so be strategic about your own self-promotion. If you want to be considered, you need to cooperate. If you want to be successful, you need to be proactive.

Chapter 7

Resiliency
The Winner's Edge

If the last few years have taught me anything it's this: The one thing you will need more than anything else—more than even proper research and interview preparation—is resiliency. Resiliency is the winner's edge in this, or any market.

"Winning is not a sometimes thing; it's an all-the-time thing. You don't win once in a while; you don't do things right once in a while; you do them right all the time. Winning is a habit," so says Vincent Lombardi, football's winningest coach.

The tech-community has seen more change in the last 12 months than in the past five years, and most of it is quite disturbing. Billion-dollar companies—gone! And not just one or two—we are talking about scores of them—bankruptcies, mergers, acquisitions, and total retrenchments. Personally, my mutual fund didn't see this coming, so I am now on the "Freedom '95 Program."

Companies are at a strategic fork in the road. One path is to hunker down, cut costs, and opt for incremental tactics. The other is to boldly make changes. To lead an industry, employers must walk the bold road. Now, more than any time in the past five years, they need the winner's edge.

Tomorrow's next generation of winners are, just as you are, busy plotting the technologies that will revolutionize the world. These are the smartest, savviest, most well-educated, and most ambitious people in the world. The more ubiquitous technology becomes, the greater a company's success depends on hiring people such as these. The tech-wreck of 2001 is a constant reminder of how seldom companies institutionalize the practice of hiring winners.

In the old days (five years ago), a talent search could be likened to a two-dimensional board game. It allowed a company a leisurely amount of time to send out requests for resumes, which could then be judged along a price/performance set of axes. It allowed for a precise, two-dimensional grid and a search process so structured and simple that a corporate HR department's less skilled staff could handle judgments.

Now, the world moves in three dimensions, at a speed more resembling a video game. It's zap or get zapped, in real time. In such a fast-paced dogfight, you need to define your missions very carefully: no wasted energy, no blurred vision. The right people are more important to companies than ever before. Don't leave your career in HR's hands.

Decision-makers are rarely looking to fill in a box on a standard employee recruitment form. They're usually looking for something more nebulous and more important. They're looking for a person who can deliver a *quality*, not a *quantity*. Remember the New Value Table? It's up to you to articulate your value every step of the way—even more so with HR people.

Resiliency: The Winner's Edge

Winners share one special quality: resiliency. As defined by the American Heritage Dictionary of the English Language:

> **re·sil·ien·cy**—The act of resiling, springing back, or rebounding; as, the resilience of a ball.

For our purposes, resiliency quite simply is the ability to take disappointment, or "take a hit" as Vince Lombardi would say, get up, and keep going.

You know full well that if you were at work and you ran into an obstacle that prevented you from launching a new product or reaching a key client for information that was going to help you be more successful on the job, you'd find a way around it or through it.

Put another way, if you found out you had the winning lottery ticket in your hand with 15 minutes left to claim the $5 million you'd find a way to do it—in a snow storm—with a leg cast!

Your job search is no different and no less lucrative (40 years × 125K = $5 million). This is your life. Take complete charge of it. Develop your plan and then start implementing it with resiliency. Parlay your series of jobs into a purposeful career.

Firsthand Experience

All around you the people who have the character and courage you admire most in life have gone through pivotal moments in their lives, times where they've pushed forward and won against all odds. Whether it was a job loss, a death in the family, a health problem, or something as "simple" as getting through college/university, you admire them for it. That experience gave them resiliency and transformed them into winners. From that moment forward winning becomes habitual for them—and their resiliency permeates everything they do.

Who Are These People and What Are Their Stories?

It's the young girl who makes the Olympic Ski Team and on her first run defects from "Mother Russia" and rises through the ranks of a technology juggernaut; the publisher who says no to "sex ads" in his newspaper and is exiled, only to be brought back to resurrect lagging sales; the housewife who puts her career aside to raise four children and builds a profitable consulting company out of necessity. These are the stories of people who have developed resiliency. You know people just like them, and you know their stories.

What Are Yours?

You've had these pivotal moments throughout your life, and you've become stronger in character because of them. While you are on your job search I suggest you write a note to yourself about them. Then, whenever you're feeling down because you're not meeting your objectives, read your note out load. Relish the moment and remember how you felt beating the odds—then get back to work!

Your Commitments to Succeed

Faced with a stagnant economy, you can pull in your horns and settle for an uninspiring routine job that doesn't challenge you *or* seize the moment and work through the problems. You know full well everything in this book makes sense. There is no silver bullet. Finding a job is a lot of work...mapping out a career even more so, but it does work—if you work at it.

Now that you are an expert job seeker I want you to

1. Define the environment that gets your motor running.
2. Detail your career history with absolute precision.
3. Define the industries of interest to you.
4. Analyze your skills and personality characteristics.
5. Draft your resume.
6. Determine your attack strategy (10 ways to find a job).
7. Draft your cover letter templates.
8. Organize your computer for highly effective and rapid response.
9. Prepare for future interviews.
10. Concentrate on market research according to your strategy.

Now, you are ready to attack these exciting opportunities!

Perry's Rules

Rule 1: Consider every minute you spend on job hunting as an excellent investment in your future.

Rule 2: Never get frustrated.

Rule 3: Dedicate sufficient time every day for market research to fill your "opportunity funnel." (I suggest at least two hours.)

Rule 4: Never compromise on any of your commitments.

Rule 5: Thank everyone you come into contact with through the process. Remember Pavla's hairdresser.

Rule 6: When you do finally land the job of your dreams remember how you felt along the way, and make a personal commitment to go out of your way to help everyone who calls you looking for help or advice.

Rule 7: Become a source for headhunters. They'll appreciate it, remember you, and reciprocate.

Rule 8: Thank your family once a week for their support. If they offer, let the kids (or someone else who cares) take you out to lunch. They need to feel they are helping.

Rule 9: Consult the New Value Table for everything you do and every job you apply to.

Rule 10: Become super-resilient.

Tomorrow

I hate "rah-rah" motivational pitches, so I am not going to give you one. There is, however, one truth I've heard over the years that I know to be true: The mind can only hold one thought at a time. Shortly after the tech-wreck of 2001 a friend reminded me that I could either complain about having no business or focus on writing more new business, but I couldn't do both, and only *I* could decide which to do.

At the time I was five months into a severe business drought. I chose to write more business, and on my 1,114th cold call my luck changed. And on my 1,115th, and 1,116th. You never know when and where your good fortune will take you. We're at the end of the book and I wish I could now give you a silver bullet solution for finding a job. But I can't. I can only give you the process and encourage you to follow it.

You now have all the tools you need to create the career you deserve. As you go through your job search, people who don't return your calls or who don't quite grasp your value will inevitably disappoint you. Keep feeding your opportunity funnel with new companies and new interviews. Learn from each encounter. Hone your pitch.

Today's economic meltdown is forging tomorrow's market leaders. Be one of them! The future is in your hands.

Carpe Diem!

Appendices

Appendix A: Career History

Take your time and make as many copies of the following form as you need. It's not a contest, nor is it a race. I want you to really detail all your experience for every job—especially the accomplishments—because you can't possibly sum up your value in two or three pages on a resume.

You will need to recall information and accomplishments that aren't on your resume but are important to a particular job opportunity on the fly in an interview. By writing out and referring to these notes before major interviews you may well be able to say to an interviewer, "this isn't on my resume *but* I think X is important to success in this role, and at Y corporation I did."

You don't want to have to think too long or too hard to recall this information later. It will be on the tip of your tongue if you spend the time now doing this exercise thoroughly.

◆ ◆ ◆

Your Career History (Full-Time and Part-Time)

EMPLOYER:_____ DATES:_____

TITLE:_____

RESPONSIBILITIES:_____

_____

_____

_____

_____

_____

ACCOMPLISHMENTS:_____

_____

_____

_____

_____

_____

──────── Appendix B: Volunteer History────────

Do not underestimate your volunteer experience. I pull more critical skills from volunteer experience when I am constructing a resume for people than I do from their regular jobs.

Spend the time here to do a good job. Take a look at the action words in Appendix C for inspiration.

◆ ◆ ◆

Volunteer History

ORGANIZATION:_____ DATES:_____

TITLE:_____

RESPONSIBILITIES:_____

_____

_____

_____

_____

_____

ACCOMPLISHMENTS:_____

_____

_____

_____

_____

_____

Appendix C: Action Words

Action-oriented words create pictures in the reader's mind. *Accelerated*, for example, implies speed, quickness, and control. The mental picture the reader gets is of a very focused effort from a hardworking dedicated employee who'll make them money, save them money, or increase their efficiency.

I also use the past tense of each verb because that implies it is/was an accomplishment. Subtle but very effective.

Use the words on the following page sparingly. Use them correctly. When in doubt, use a dictionary to ensure that you use the word correctly.

◆ ◆ ◆

accelerated	composed	exchanged	led	publicized
achieved	conceptualized	expanded	located	published
acquired	concluded	expedited	maintained	purchased
activated	condensed	explained	managed	pursued
adapted	conducted	exposed	marketed	qualified
added	consolidated	extended	maximized	quantified
adjusted	constructed	extracted	measured	quoted
adopted	consulted	facilitated	mediated	raised
advanced	contracted	financed	mentored	rated
advertised	contributed	focused	merged	recommended
advised	controlled	forecasted	met	reconciled
aided	converted	formalized	minimized	recorded
allocated	convinced	formed	modernized	recovered
altered	coordinated	founded	modified	recruited
analyzed	corrected	furnished	monitored	rectified
answered	created	gained	motivated	redesigned
anticipated	critiqued	gathered	multiplied	reduced
applied	cultivated	gave	negotiated	refined
appointed	cut	generated	nurtured	regained
approved	debugged	granted	obtained	regulated
arranged	decentralized	guided	offered	reinforced
assessed	decreased	handled	opened	rejected
assigned	defined	headed	operated	remedied
attained	delegated	hired	orchestrated	remodeled
attracted	demonstrated	hosted	ordered	renewed
audited	designed	identified	organized	reorganized
authored	determined	implemented	outlined	repaired
authorized	developed	improved	overhauled	replaced
automated	devised	increased	oversaw	reported
balanced	directed	influenced	paid	represented
bargained	disclosed	informed	perfected	requested
bought	discovered	initiated	performed	researched
broadened	distributed	inspected	persuaded	resolved
built	documented	installed	pioneered	restored
calculated	doubled	instigated	placed	restructured
centralized	drafted	instituted	planned	revamped
chaired	earned	instructed	positioned	revealed
championed	edited	integrated	prepared	reversed
changed	eliminated	interpreted	presented	reviewed
charted	employed	interviewed	presided	revised
chose	enabled	introduced	prevented	revitalized
clarified	encouraged	invented	priced	rewarded
classified	enforced	invested	prioritized	saved
cleared	engaged	investigated	processed	scheduled
closed	engineered	invited	procured	screened
collaborated	enhanced	isolated	produced	secured
collected	enlarged	issued	programmed	selected
combined	enlisted	joined	projected	sent
compared	established	judged	promoted	separated
compiled	evaluated	launched	proposed	served
completed	examined	lectured	proved	set

Appendix D: Value-Based Resume

Here is my personal resume, circa 1990. I'm certain your accomplishments will be larger/greater/more than mine, so no laughing. It's the structure of the document as much as the content that counts.

The style has become very popular. I see examples of it all the time from California to New York. Many professional resume-writers now use it.

Normally I suggest you put your education at the end—*except* when you have a degree or accreditation that you know the company is looking for. Most "technical resume readers" screen resumes for education first, contrary to what all the popular books tell you.

When they pick up your resume, if education is not first they'll flip to the last page to view it. The problem is that they may not find it and discard you immediately or start reading your resume from back to front, in which case your carefully laid out presentation loses all its effect.

◆ ◆ ◆

DAVID PERRY
451 Daly Avenue
Ottawa, Ontario K1N 6H6
Telephone: 613-555-5555
davidperry@website.com

ACADEMIC EDUCATION

McGill University, Montreal, Quebec
Bachelor of Arts, Economics (1982)

SUMMARY

Over ten years progressive experience in human resources, marketing and consulting, business development, product development, project management and sales to local and national private and governmental organizations. Leadership ability developed in the successful formation, focus and guidance of new ventures.

CURRENT RESPONSIBILITIES

As Managing Partner of a Human Resources Consulting Firm my primary responsibilities include management and operation of the day to day affairs of the firm including: development of new business activities, client presentations, hands-on projects management, research, detailed report writing, as well as supervising subcontractors.

Fully versed with MS-DOS/Windows and numerous database, project management and word processing software packages. Have completed professional development seminars in Leading Edge Marketing Strategies, Effective Time Management, Personnel Supervision, Leader Effectiveness Training, etc.

My interpersonal skills have been a key component in the successful development of new business activities.

Contact Info:
You have two ways to connect with me mail or telephone.

Education:
I broke the rules and put education first because I wanted to highlight my negotiation skills (skills I aquired studying industrial relations). Yes, normally education is last.

Summary:
This is a two- to three-line, high level overview of my career. It's a snap-shot not a portrait. It's concise. It's relevant to the job I am pursuing. Yours needs to be relevant too.

Current Responsibilities:
This is a two- to three-paragraph summary of my current job and the skills I use to perform it that are relevant to the job I am pursuing.

This is where I single out my greatest strengths ("my interpersonal skills").

CAREER HISTORY

1988 to   Principal and Managing Partner
          Perry-Martel International Inc., Ottawa, Ontario
          • Executive Search & Strategic Human Resources Consulting

1986-88   Senior Account Executive
          ConsulPro, Ottawa, Ontario
          • Executive Search

1986   Account Executive
          Harrington Personnel, Ottawa, Ontario
          • Personnel Agency

1985-86   Client Services Officer
          Royal Trust, Ottawa, Ontario
          • Financial Institution

1982-85   Senior Manager (promoted five times in three years).
          Consumers Distributing, Ottawa, Ontario
          • Retail Management

MAJOR ACCOMPLISHMENTS

Consulting
• started and successfully developed a human resources consulting firm
(Sept.1988), which specializes in executive search and strategic human
resources planning. Firm has received national attention in our specific
market niche and increased revenues by at least 40% each year since 1988.

Business Development
• planned and implemented the market entry strategy for a new consulting
division, including: primary market research, target market definition, pricing
and service literature.

• led proactive business development campaign in Toronto and Montreal
which resulted in the acquisition of four major blue chip accounts from direct
competitors: subsequent revenues exceeded $250k and accounted for 37%
of the firm's annual revenue.

• recommended and initiated the formation of a new profit center resulting in
complete domination of the local Ottawa market and a strong national presence.

• conceived, developed, tested and brought to market two proprietary groups
of value-added services that put Perry-Martel on the leading edge of
executive search and strategic human resources planning.

People Management
• established a unique staff training program selecting and training 5
managers and more than two hundred staff over a three year period with a
99% retention rate.

• motivated and retrained staff in four turnaround situations in Nepean,
Kanata, and Ottawa, Ontario.

• reduced extreme turnover and absenteeism rates as well as inventory errors
to negligible amounts (turnover reduced from 97% to 0% on staff of 45), (shrink
reduced from $660,000.00/year to $126.00/year on sales of $4.1 million dollars).

Career History:
From my Career History the reader can tell I'm in the HR field, have 10 years experience, and have management experience.

Here are the functional groups of accomplishments (instead of skills) I chose to highlight.

The ordering of the Accomplishments is in response to what I perceived to be his interests:

Consulting

Business Development

People Management

Leadership
• successfully demonstrated leadership by motivating apathetic staff to be productive, resourceful and efficient, resulting in winning 3 National and 39 District sales contests.

• instructed and evaluated 45 Reserve officers in leadership, methods of instruction, problem solving, effective communications, administration and military law.

• entirely responsible for the training and welfare of 30 of Canada's top army cadets at the Arctic Indoctrination Course in Inuvik and Tuktoyuktuk Northwest Territories, including teaching courses on Advanced Leadership and Arctic & Rugged Terrain Survival Techniques.

Management of Assignments
• developed a proprietary project management system that gives Perry-Martel a distinct competitive advantage when it comes to managing and delivering cost effective value-added consulting services.

• managed 15-20 major projects per year since 1988 on a local and national status, bringing them all in on time and within budget.

Sales
• increased retail sales revenues from $1.7 million to $2.5 million over a five month period by dramatically improving customer service, personnel training, and materials management in one store, which led to another turnaround situation where sales were taken from $2.3 million actual ($3.3 budgeted) to $4.1 million in 11 months

• increased the branch office assets of a bank (client deposits) by $15 million in a six month period through target marketing and designing a selective referral program.

Marketing & Public Relations
• designed and implemented a "no-cost-to-us" strategy of raising corporate profile and highlighting the company's strengths through a public relations and marketing campaign that included major local and national television, radio, newspaper and magazine coverage. Direct business resulting from the campaign is worth approximately $400,000.00 and growing.

• Recognition regularly quoted by the media on local, national, and international employment issues.

EXTRA CURRICULAR ACTIVITIES (concurrent with education and full-time work)

Military Training
Jan.-Apr.'82 Captain Qualifying Course, Royal Canadian Army Cadet Instructors List
• successfully completed course and promoted to Captain.

Jun.-Aug.'80 Summer Officer Training Program, (R.C.A.C.I.L.)
• graduated 1st. of 126 officer candidates.
• selected to teach winter session at School of Instructional Technique.
* awarded SWORD OF HONOR by Brigadier General J.B. Dabros.

Leadership

Management of Assignments

Sales

Each section builds toward the conclusion I want him to reach—that I am qualified to handle his project—despite my youth.

I brought in the military training because it shows I am more than a textbook consultant. It also demonstrated that I walk the talk on working hard.

Work History (part-time: evenings and weekends)

1982-84 Commanding Officer - Canadian Grenadier Guards Cadet Corps, Montreal.
     • supervised team of 5 officers, monitored training of 55 cadets, administration, public relations and fund raising.

Jun.-Aug.'81 Platoon Commander - Longue Pointe, Quebec.

Jun.-Aug.'82 Platoon Commander - Arctic Indoctrination Course. Tuktoyuktuk and Inuvik, North West Territories.

Feb.-Apr.81 Instructor - School of Instructional Technique. Montreal
     • instructed 60 Reserve Officers in; leadership, methods of instruction, problem-solving, effective communications, administration, and military law.

1973-1981 Training Officer /Supply Officer /Civilian Instructor/Cadet, (GreenfieldPark & St. Jean, P.Q., Borden, Ontario)

AWARDS & HONORS

1986:   Rookie of the Year, (not awarded for previous 3 years) ConsulPro, Ottawa, Ont.

1982-85: Won 3 National & 39 District Sales Awards, Consumers Distributing, Ottawa.

1982:   Nominated for Rhodes Scholar, McGill University, Montreal, Quebec.

1980:   Sword of Honor, Summer Officer Training Program, Montreal, Quebec.

1979:   Head Boy, Base Borden Collegiate Institute, Borden, Ontario.

1978:   B'nai Brith Citizenship Award, Centennial High School, GreenfieldPark, Quebec.

1976-79: High School activities included; President of Student Council, Year Book Advertising Editor, Grad Committee Chairman, Cadet Commander, Officer Cadet, Civilian Instructor.

PUBLICATIONS

"S.M.A.R.T.", (Registered Trademark, 1992).
"S.H.A.R.P.", (Registered Trademark, 1992).
(Strategic Human Resources Planning), Life-cycle Planning System. Perry-Martel Int. Inc.(1991). (Strategic Marketing and Recruiting Techniques),
Search System, Perry-Martel Int. Inc.. (1990).

He could clearly see I had worked my way through university.

Through the Awards section he could see that success was a re-occurring theme throughout my life.

Normally I would suggest you leave out high school awards but in this case I needed it to show that I had a history of success in everything I touched.

——Appendix E: "Before" and "After" Resumes——

These are typical resume makeovers you can do for yourself. Have your resume read by someone who doesn't know you (other than the guy at the local coffee shop) to see if he or she can tell what you do for a living. If the reader doesn't get it, chances are good that your intended target won't understand it either. Of course, if he or she says, "Wow—you're some kinda technical geek, aren't you?" or "Hey man, do you know Bill Gates?," it's probably fine, but have it checked by a colleague anyway.

◆ ◆ ◆

Engineering Resume Before

Career Objective

Sales / Applications Engineer:
To use my technical and business skills in combination with my sales and international experience to help grow an established company as a Sales or Applications Engineer.

Professional Skills

- Experienced in acquiring and managing clients.
- Strong technical background.
- Well-developed oral and written communication skills.
- International experience.
- Experienced teaching IT personnel and taught university level tutorials.
- Multiple programming languages including: HTML, Visual Basic, C, C++, TCL/TK, MS-DOS batch files, SQL, various assemblers, Perl, Pascal, Dbase and numerous scripting languages.
- Worked on, maintained and installed multiple computer platforms, including: Win95/98/NT and Linux.
- Communication protocols/systems: TCP/IP and various Internet servers (e.g. Email, FTP, Proxies, Web, etc.).
- Digital hardware design.

Significant Contributions
Business Development:

- Established an Internet consulting firm.
- Founded Verisim, Inc. with 3 technical people and sold the technology.
- Supported 4 partners and 3 graphics people at Verisim, Inc. by acquiring business through my Internet consulting firm.
- Sought out new clients and managed accounts.

Management:

- Spent two years running an IT department in Malawi Africa as part of a World Bank initiative.
- Managed small teams up to 7 people on multiple projects.
- Supervised and trained users in various OS environments.

Technical:

- Developed, deployed and implemented a network supporting 200 people and 2000+ projects on 90 computers across 13 sites.
- Member of a team that wrote a new type of continuous time, massively multiplayer, Internet game engine.
- Conceived multi-level core technology able to turn games out very quickly with a minimum of rewriting.

Professional Experience

MIS/IT Specialist, 1999 – 2001, MASAF, Malawi, Africa,

- Revamped and improved IT infrastructure, managed IT department, oversaw the development of an MIS System, and trained both technical and managerial IT staff.

Vice President Marketing and Finance, 1994 – 1998, Verisim, Inc., Nepean, ON

- Founded the company (7 people), wrote the business plan, acquired financing, performed market research, planned business direction, worked on product development and *sold the company.*

Engineering Resume After

Since graduating as a Computer Engineer from the University of Waterloo seven years ago, I have:
- Started an Internet consulting firm.
- Founded a massively multi-player, continuous time, on-line gaming company *and sold it.*
- Spent two years running an IT department in Malawi Africa as part of a World Bank initiative.

I have just recently returned to Canada and am now looking for employment in the Ottawa area. I believe the best use of my skills would involve working with engineering in conjunction with customers and/or strategic partners, perhaps as a Systems Engineer or Product Manager.

Employment History:

MASAF, Malawi, Africa, MIS/IT Specialist, 1999–2001
Revamped and improved IT infrastructure and oversaw the development of an MIS System.

ABC Net, Inc., Nepean, ON, Vice President Marketing and Finance, 1994–1998
Founded the company (7 people), wrote the business plan, acquired financing, performed market research, planned business direction, worked on product development and *sold the company.*

David Harry & Associates, Ottawa, ON, Principal, 1994–1998
Internet consulting business that developed Internet strategies for small businesses.

Coop Student, 1990–1993
Work terms included four, four-month terms at Nortel (software programming to digital hardware design).

Significant Contributions:

Technical:
- Starting from ground zero, developed, deployed and implemented a network supporting 200 people and 2000+ projects on 90 computers across 5 sites.
- Wrote a new type of continuous time, massively multiplayer, Internet game engine.
- Conceived multi-level core technology able to turn games out very quickly with minimum rewriting.

Business Development:
- Supported 4 partners and 3 graphics people at ABC Net by acquiring business through D. Perry & Associates.
- Founded ABC Net with 3 technical people and sold the technology to a major game house.
- Sought out new clients and managed accounts including ones from the CATA Alliance.

Management:
- Managed small teams up to 7 people on multiple projects.
- Supervised and trained users in various OS environments.

Technical Skills:

Programming Languages Most Recently Used: HTML, C++, TCL/TK, MS-DOS Batch Files
O/S installed and maintained: Win95, Win98, WinNT Server /worked with: Linux, Win9x, WinNT
Communication Protocols/Systems: TCP/IP and various Internet servers

Education:

University of Waterloo, Waterloo, ON *B.S., Computer Engineering,* **1994**
Courses included: Entrepreneurship, Software Development, Control Systems and Digital Design.

Computer Operator Before

B. Person
451 Daly Avenue
New York, New York 98754
Telephone: 212-555-1212

WORK EXPERIENCE:

April 1990 to Now

Computer operator
For the ACS-COOP using different systems: Geac, IBM Mainframe (4381, 9672), Stratus, Series 1, LAN, and personal computer equipment

IN ADDITION to performing the duties and maintaining the responsibilities of computer functions, following are additional tasks which have been assigned to me and which I continue to perform in a fully satisfactory manner in my present position:
- I automate LOTUS 123 files for statistical purposes on monthly reports
- I serve as backup to the supervisors' absence and as their right hand
- I work fluently with communication equipment in different technologies and settings
- I developed and continue to maintain cable network for the Magicbanc network
- I clean up the computer room by recycling hardware, thus saving cost to the ACS-COOP
- In conjunction with the Bank of Montreal, I ensure dial backup systems availability
- I am the combination custodian for computer room vault
- Train operator staff on new duties and ensure daily production schedule is maintained to standard

AS WELL, I have had other projects, now completed, which I am proud of:
- I developed backup of the "forms design" PC
- In 1993-94, I acted as supervisor for over one year in the computer room

November 1989 to April 1990
Automated teller clerk
I worked full time in the automated teller department
- verifying deposits
- updating General Ledgers
- ordering ATM cards for members
- handling calls from members for lost and stolen cards
- having the responsibility of one of the two combinations for the vaults

CONTINUING EDUCATION:

1998	- C++ computer programming course at Algonquin College
1996	- A+ certification service technician for computer hardware with specialty in DOS and Windows environments
1995	- COBOL programming
1995	- Service plus offered by the CS-COOP
1991	- SRA courses offered by the CS-COOP (JES2 system control, JES2 job and device control)

(MVS operator training, MVS concepts and facilities)

SELF-LEARNED
- OS/2, DOS, WINDOWS, experience at home and work.
- Communication courses 1984-1985 Edmonton Alberta
- Bookkeeping basic course at NAIT in 1981
- Private pilot's license summer 1977

trained all new people:
- trained quickly on series 1 system for atm network to integrate others on how to use the system effectively.
- fully trained after two months on GEAC 8000 to operate the banks financial system, there was no formal training system available so I filled in to train new computer operators.
- Was successful in replacing supervisor to train new and existing employees for 4 years before starting new projects.

designed decision support process for senior management:
- created solutions for disabled atm machines and ways to report useful information.
- troubleshoot with outside agencies to fix our hardware/software breakdowns.
- worked with programmers to update information to errors on unit and messages given on screen
- maintained services to all departments and computer hardware on daily basis
- recorded daily activities as to inform other crews on shift and update superiors problems/resolutions

reorganized access matrix to save $
- helped organize layout of hardware on computer room floor for maximum efficiency/cost
- reduced from 7 communication controllers down to 3 saving maintenance costs
- reduced downtime to users on the system in may ways
- gained useful understanding to the networks and systems to assist in solutions for problems.

reorganized cabling topology to maximize efficiency
- cleaned wiring mess to 250 dumb terminals
- created schematics to track each terminal
- grouped cables and hardware for best use kept it neat for future growth
- worked with contractors for location and documentation of access terminals

initiated ongoing preventative maintenance program for hardware
- organized spare stock of equipment and kept lists of serial numbers where pieces are
- cleaned hardware for longer terminal life
- worked with large laser printer, automated teller machines, mainframe terminals

established guidelines for a shift 24/7 work environment
- changed day/night shift from 12 hour shifts to combination of 8 and 12 hour shifts for easier resolution to holidays and sick leave
- made department much more efficient to user needs
- reduction in time outages for critical systems
- optimized supervisors time on working with employee's
- established recycling program saving $10,000 for data cartridges
- worked with outside agencies to recycle hardware
- expanded our database storage capacity for future growth
- ensured growth when budget was exhausted for our plans
- worked with system applications to ensure quality of tapes and reliability

5 years no sick leave
- keen interest in job growth kept my interest active
- committed to others in training programs in our area
- worked in a 7/24 job schedule day and night rotation
- took on responsibilities outside of my duties for incentive to grow

9½ years same company, took buyout package
- turnover changed dynamic of department and schedule of duties
- worked at different levels with several departments to streamline our activities with their demands
- changed computer systems as technology grew and demand changed
- more automation changed requirements and knowledge of employees duties

streamline the payday and new product process
- worked with supervisors to establish automated process of deposit and withdrawal of money from members accounts
- updated system as more products were added
- recorded documentation on how process worked in our area various shifts and who was responsible for jobs failing

acted/created the help desk support function
- worked with PC department as a support form network to mainframe problems
- if a PC had a problem would resolve outage internally giving a quick turnaround
- turned position into backup of help desk problems

software used
- DOS; Windows 3.1,95; OS/2 2.1, warp; MVS (mainframe) deal; VOS (midrange); IBM series 1 Cobol, assembler; GEAC 8000 (mainframe)' C++ and COBOL programming Lotus notes, work pro, lotus 1-2-3, freelance graphics, lotus approach, Harvard graphics, WordPerfect, Netscape navigator, Microsoft explorer, Excel, Quicken for Windows, several anti virus software (thunderbyte, mcaffey, Norton, other Internet versions)

Computer Operator After

SUMMARY

Nine years' progressive experience as a Computer Operator and ATM trouble-shooter for a medium sized financial institution.

CURRENT RESPONSIBILITIES

As the Computer Operator for the ACS-Coop Head Office in Ottawa (1990–1999), I ensured the integrity of the internal network of 250 terminals and PCs as well as the external Automated Teller Network in sixteen locations spanning from Ottawa, west to Toronto, and north to North Bay.

I was responsible for the day-to-day operations of the Mainframe and the Magicbanc network. In conjunction with the Bank of Montreal I ensured dial backup systems availability.

I am fully versed with MS-DOS/Windows and numerous database, project management, and word processing software packages. I have completed professional development seminars in customer service, bookkeeping, and communications. I have also successfully completed ACS-COOP funded courses in: JES2 system control & job and device control, as well as MVS operator training.

My easygoing nature has been a key component in the successful operation and maintenance of a complicated, widely dispersed technology environment on a tight budget.

CAREER HISTORY

April 1990–May 1999   ACS-COOP, Ottawa head Office
Computer Operator *(opted for buy-out package in May 1999)*

Nov. 1989–April 1990   Automated Teller Clerk

SUPPLEMENTAL CONTRIBUTIONS

- established training procedures and trained new operators on the GEAC 8000;
- designed decision support process for senior management;
- reorganized access matrix to save $$$;
- reorganized cabling topology to maximize efficiency;
- initiated ongoing preventative maintenance program for hardware;
- established guidelines for a 24/7 shift work environment;
- streamlined the payday and new product process; and
- created the help desk support function

TECHNOLOGY PROFICIENCIES

Hardware:
Geac 8000; IBM Mainframe (4381, 9672); Stratus; Series 1; LAN; and various PCs

Software:
DOS; OS/2; Windows 3.1,95; OS/2 2.1; Warp; MVS (mainframe) deal; VOS (midrange); IBM series 1 Cobol; Assembler; GEAC 8000 (mainframe); C++ and COBOL programming; Lotus Notes; Work Pro; Lotus 1-2-3l Freelance Graphics; Lotus Approach; Harvard Graphics; WordPerfect; Netscape Navigator; Microsoft Explorer; Excel; Quicken for Windows; several anti virus software programs including: Thunderbyte; McAffey; Norton; and other Internet versions

VP Marketing Before

Kathy Star

Xxxxxxxxxx
xxxxxxxxxxx
Home: xxxxxxxxxxx   Fax: xxxxxxxxxxx   Cell: xxxxxxxx
e-mail: xxxxxxxxxxxx

Career Profile

A seasoned strategic thinker with a proven track record in business planning, empirical analysis, negotiation and business development. Experienced officer of publicly traded companies dealing with investors, analysts, and public relations contacts. Fully prepared and motivated to take on the role of President/CEO of an emerging Internet software company.

Senior management experience in all aspects of marketing including product management, competitive research, channel development, Internet-based marketing, international marketing and marketing communications. Proven ability to manage remote international offices and geographic specific marketing initiatives. An adept leader with a demonstrated record of successful mentoring, team-building, and objective achievement.

Career History

AUG'01 – PRESENT   COMPANY
Start-up maker of XML-based data migration technology. Privately-held company.

President
Developed vision and business plan that brought several VCs into discussion (due diligence continues). Executed new technology roadmap and delivered first product to market on time. Helped establish three major potential lead customers (negotiations are ongoing).

OCT'00 – AUG'01   COMPANY
Leading maker of e-business analytic and reporting software. Privately-held company. Part of 7-member senior management team. Signing officer of Company.

Vice President, Marketing
Responsible for corporate and product marketing strategies and tactics, including consulting services, business development, channels and channel partners, strategic marketing, product management, communications, public relations, branding and positioning. Defined and launched two new products and helped steer Company into a new vertical market. Successfully grew sales pipeline from $100,000 to almost $7 million USD.

OCT'99 – SEPT'00   COMPANY
Leading supplier of ebusiness access and integration software solutions. Company is a public company (NASDAQ: xxxxx) with revenues of $150 million

in 1999 and over 800 employees. Ranked in the top 50 software companies in the world. Part of 10 member senior team leading Company corporate worldwide. Signing officer of Company Canada.

VICE PRESIDENT, MARKETING, eSOLUTIONS GROUP.

Responsible for worldwide marketing strategy and branding of Company's new ebusiness product suite, ABC Integrator, and underlying ebusiness marketing direction. Joint responsibility for Company corporate marketing efforts with VP, Marketing of eAccess Group. Direct accountability includes product management, product marketing, marketing programs, field support, marketing communications, public relations, web marketing, branding and positioning.

Managed eSolutions Group with two other senior VPs for three months and built new business and operations plan for FY 2000 to increase ABC revenue by 100% and position ABC as the go-forward ebusiness product for Company over the next five years. Executed on plan through 2000 and helped drive an increase of 45% in Q3 00 revenues over the previous comparable quarter.

Nov '97 – Oct '99   COMPANY
Leading supplier of web-based extranet software and applications. $22 million in revenue. Publicly traded on NASDAQ: SIMW.

VICE PRESIDENT, MARKETING

Responsible for full marketing strategy and tactical plan for company and products, including flagship Internet product. Grew Software B revenue by over 400% and increased Software B revenue contribution from 15% to 60% of total in two years. Responsible for a marketing budget of $3 million that includes North American and European marketing initiatives and staff. Manage worldwide marketing organization including European offices in England and Belgium.

Played an important role in direct communications with industry analysts, investors and financial analysts. Strategically positioned AirTite via the XXX extranet product to win the leading Crossedroads award for best product in category in 1998.

Selected to be part of a 4-member team responsible for successfully integrating Company after the purchase of the company by Company of California.

Jan. '97- Nov '97   COMPANY
Fast-paced start-up Internet-based network management software company.

VICE PRESIDENT, MARKETING

Part of team that secured $2.2M venture financing. Responsible for marketing budget of $1.0M. Developed marketing plan that played a key role in acquiring funding. Company included in *Financial Best* top 25 Canadian software companies.

Developed and implemented strategic marketing plan, including product, promotion and distribution. Successfully launched three new products into fast changing Internet market resulting in significant positive media coverage, increased evaluation downloads of product and increased sales.

Developed Internet marketing strategy that increased web site traffic by 300%, resulting in 80% of total sales coming from the Internet.

Responsible for the development of collateral, packaging, and other support materials: established positioning, developed creative strategy and managed production. Developed new corporate identity. Conducted highly successful PR and media relation campaign resulting in lead stories in major industry publications. Responsible for advertising program to support marketing strategy, including print and electronic advertising.

1995-1997

COMPANY
Canada's mobile satellite telecommunications company, an ABC subsidiary.

DIRECTOR, DATA AND VOICE SERVICES.
Represented voice and data services to Board as part of senior management team. Championed market introduction of data services.

Developed business models that resulted in manufacturers developing new data hardware for use on network.

Successfully convinced the organization to bypass existing service provider model for data services. Identified and implemented appropriate distribution model. Established correct pricing resulting in lead customers.

Responsible for all business aspects of data services unit, including marketing, sales, product management, marketing communications, business plans, distribution, channel models, profitability and leadership.

Successfully negotiated and signed lead customers before service was available. Convinced service providers they had a role to play.

Turned a troubled team into a dynamic and motivated group committed to team goals. Provided leadership to product management group. Established strong and positive relationship between product managers and engineering

DIRECTOR, STRATEGIC MARKETING.
Developed company's strategic business plan, including 10-year view and new market and service opportunities.

Developed strategy that resulted in underwriters' interest in an IPO.

Conducted business analysis of potential purchase of US company that would have allowed the extension services into the US and eliminated future threats from US-based competitors.

Negotiated MOU with major satellite manufacturer to develop second-generation satellite.

Responsible for strategic alliances for mobile satellite business including companies such as Client and Client.

Responsible for market intelligence on worldwide satellite and telecom business activities and trends.

1992-1995   Company
Number one selling hearing aid manufacturer in Canada.

MARKETING DIRECTOR
Led all aspects of marketing activities including, product management, promotion, pricing, marketing communications, media and public relations, channel management and support of international distributors. Managed $2M budget.

Established a market-driven environment in an engineering organization. Created a strong 14-person marketing team that included product managers, writers, designers and software developers.

Introduced Unitron's most successful next-generation products. Reestablished Unitron's leadership in children's market by launching the highly successful Kids Klub campaign.

1989-1992   COMPANY
Consulting company providing marketing and marketing communications services.

PRINCIPAL.
Services included writing, design, print brokering, marketing plans, technical communications, press releases and speeches. Specialized in high-technology software and hardware, particularly geographic information systems.

Clients included Client

1988-1990   COMPANY
Maker of the world's leading geographical information systems (GIS) software.

MARKETING SERVICES MANAGER.
Created promotional strategies, targeted potential markets, produced marketing material for top GIS vendor and software product in Canada. Handled technical communications training requirements and proposals. Led RFP response team.

Professional Activities

Former Member of Board of Directors of the ABC Software Resources Council
Member of the Board of Directors of Texasec.
Member of the Board of inmedia Public Relations.

Education

Ph.D.   University of Toronto, 1987
M.Phil.   University of Waterloo, 1982
M.A.   University of Waterloo, 1980
Hon. B.A.   University of Waterloo, 1978
Successfully completed various courses and seminars on communications, marketing, branding and product management offered by institutions such as York University and Queens University as well as recognized business education companies.

VP Marketing After

Kathy Star

cat◆a◆lyst – an agent that provokes or speeds significant change or action

SUMMARY
Over 20 years' experience as a leader in both technical and business aspects of the technology industry, with a demonstrated ability to manage and motivate people, create commercially successful products, and develop and execute strategies that result in marketplace success. My most recent successes include:
- Developed vision and bus plan that brought VCs to the table (due diligence continues)
- Executed new technology roadmap and delivered product to market on time
- Drove a pipeline increase from $100,000 to $7 million within 8 months
- Defined and launched two new products into a new vertical.
- Designed and executed client acquisition plan resulting in 45% increase in revenue.
- Ignited new product revenue 400%.
- Positioned company to win the "Crossroads – Best product in category" award.

CAREER DRIVER
President/CEO of an emerging Internet company that has an aggressive mind-set.

SPECIAL SKILLS
My start-up experience has honed these skills:
- Management - optimizing people and finances to meet objectives
- Budget maximization - for effective use of finite resources
- Creative concepts - branding, e-mail marketing, partnerships
- Experimentation - relentless probing for new revenue approaches
- Expressive clarity - industry commentary, outreach, Web writing
- Conceptualization - analyzing structures and policies

EMPLOYMENT HISTORY

Company, *President (08/01–present)*

◆strategy	◆communications	◆investor strategy
◆marketing	◆sales	◆business development

Company, MIS/IT Specialist—Ottawa, ON (10/00–08/01)
Company, VP Marketing—Ottawa, ON (11/97–10/00)
Company, VP Marketing—Ottawa, ON (01/97–11/97)
Company, Director Data & Voice Services—Toronto, ON (95–97)
Company, Marketing Director—Toronto, ON (92–95)
Company, Founding Partner—Toronto, ON (89–92)
Company, Marketing Services Manager—Waterloo, ON (88–90)

EDUCATION
Ph.D.   University of Toronto—1987
M.Phil.   University of Waterloo—1982, M.A.—1980, Hon. B.A.—1978

BOARD OF DIRECTORS
- Software Human Resource Council
- xxxxxxx Public Relations

John Walton, EET, B.A., PMP

xxxxxxxxxx	Tel.: xxxxxxxxx
xxxxxxxxxxxxxx	Cell : xxxxxxxxxxx

PROFILE

John Walton is a PMP with progressive and diverse experience in the development, enhancement and accentuation of business leadership opportunities. A proven and time-tested background in creating business growth through portfolio restructuring, product and platform developments, and personnel initiatives makes him well suited to any corporate enterprise. His personal drive and dedication to the business, technical and human components in project development and management provides consistent and superior project results. His leadership of large and small teams is recognized as exceptional. The team environments created under his direction inspire members to achieve both personal and company objectives that translate into cost effective business solutions and exceptional client relationships.

PROFESSIONAL DEVELOPMENT

Project Management; Time and Team Management; Performance Management; Problem Solving and Decision Making; Influence Building, Using & Sustaining; Selling Ideas; Creating Customer Focus; Negotiation Skills / Listening; Defect Prevention and Root Cause Analysis; CSA/UL & ISO/TL 9000 Certification; WHMIS (Safety) training

CAREER SUMMARY

ABC NETWORKS (ABC TELECOM) LIMITED   *1983 –2001*
FPGA/Hardware Designer   Madonna Packet Switching, Ottawa   December 1999 to May 2001

This was a hands-on junior engineering position accepted with ABC in Ottawa due to a lack of Project Management opportunities within the Greater Toronto or Kitchener-Waterloo regions in 1999, and the hope that ABC's recovery would be in the short-term.

Engineering duties:

- Learned, then coded (Verilog) and synthesized (Synplicity) the RTL for the Ethernet to SONET interface FPGA on the OPTera shelf provisionable products.
- Wrote the Design Specification and Test Plan documents for the products that

Project Management duties:

- Coordinated and led both informal and scheduled development team meetings to introduce STS-12 and STS-48 interfaces on provisionable cards in the OPTera Metro 3000 platform.
- Interfaced with hardware development (primary and satellite teams), software development, product integration, lab support and PLM primes to ensure project communications and schedules were kept up to date.
- Worked with my functional manager to create the project plan in MS Project.
- Acted as the Subject Matter Expert to Raleigh, NC, USA engineers for and to maintain the I/O and peripherals portions of the DMS Family product system. *Public Carrier Network technical and R&D operations in Canada were closed 4Q'99 and the responsibilities were transferred to the Raleigh, NC, USA location.*

Results: Created ABC's first OC3/OC12 FPGA that was functional across multiple SONET-Ethernet interface "OPTera Metro" products. This design was based on the FPGA was a single OC3 provisionable circuit card.

January 1987 to November 1999

Sr. Projects Leader, ABC, Technology (R&D) Organization, 1999–'94

Projects Leader, Technology (R&D) Organization, 1993–'90

Task Force Leader, ABC Telecom, Quality – Customer Satisfaction, 1989–'87

Reported to various levels of management (Directors, Sr. Staff Managers, and direct-report staff managers) over this period based on who the project sponsor was, and where the business portfolio for the main portion of the project resided. **All** projects surpassed defined objectives, and were increasingly complex in technical innovation and strategic scope over the years. Several projects were simultaneously managed, and throughout the entire project.

- Researched and developed new business project opportunities to achieve strategic business objectives through new products, cost improvement initiatives, relational database and process developments, Six Sigma methodologies, Employee and Customer Satisfaction, and Portfolio and Staff development initiatives.
- Wrote Opening and Closing business cases and supporting documentation based on validated cost factors, technical advance, and customer input to gain authorizations and project funding, and monitored project budgets throughout project life cycle.
- When required as part of the project: Represented ABC at customer meetings related to project kick-offs, progress and objectives. Coordinated and developed 3rd party contractual agreements through the ABC Legal department. Maintained supplier/ contractor relationships throughout project life cycles.
- Managed (staff authority) project teams up to 30 full and part-time members, and in complexity from solely engineering to multi-site, multi-organizational and cross- functional staff members. Scheduled and led team and milestone and post-close-out meetings to ensure continued project "buy-in" and synergy among team members, senior management and project sponsor(s). Maintained Work Breakdown Structures using MS Project. Projects managed based on PMI methodologies.
- Communicated project accomplishments and strategies to senior management and on a per project basis at customer locations across Canada and U.S. through reports and on- site presentations. Generated and presented progress reports/charts to project stakeholder management primes, and at quarterly Operational Status (OPS) Review meetings. OPS members were normally ABC Business Unit Vice-Presidents, Directors and Sr. Staff Managers.
- Technical liaison to Manufacturing, Test Engineering, Change Management, and Sales and Customer Support groups for mature and active products within my portfolio.
- Led group (Brampton Lab) training sessions to introduce new work-related and employee satisfaction processes. Assisted ABC Training staff in the development of new courseware for new products.
- Participated in annual performance reviews (MFA–"Managing For Achievement") of project staff members when their functional managers requested my input/involvement.
- Mentored technical and non-technical staff as a participant in ABC's "Coaching Network" program.

Projects:

- Product Robustness Vision 2000 (Y2K) Initiative: Improved DMS100-Family product line integrity and robustness by 60%, achieved a cost improvement of $30M, and instituted several corporate and organizational processes to ensure continued high performance.
- Portfolio Management: Product portfolio reduced by 35%, and immediate savings of $7.9M.
- 3 Design Methodology projects for: DFT (Design For Testability), DFM (Design For Manufacturability), and DPP/RCA (Defect Prevention Process / Root Cause Analysis)
- 3 Corporate Processes for product life cycle management:
  - ➤ MD (Manufacture-Discontinue) and
  - ➤ EOL (End-Of-Life)
  - ➤ Portfolio Rationalization
- 2 R&D Tax Credit packages to Value Engineering for Revenue Canada
- ISO9000 Certification of the Brampton Labs organization
- Employee Satisfaction (E-Sat): Developed and administered survey package. Analyzed data and presented results and recommendations to senior management team. E-Sat was up 46% the first year for the 200+ member organization.
- 29 New Products Development & Introduction: Combined hardware and software projects on ABC's Digital Switching I/O, Peripherals and Datapath product platforms with budgets up to $7.5M.
- RDBMS (Oracle) Product Administration System / Internal Engineering Database: Liaison between the IS/IT and Engineering groups to ensure usability and technical aspects complied with user requirements. Negotiated and scheduled demonstrations, training, pre-release controlled user testing, reported on findings from sampled users, and ensured resolution to all issues.
- RDBMS (Oracle) Quality Reporting Information System: Redesign of Customer Return Tag to improve usability (~75% up from <20%), IS/IT Development to Repair Centre operation (Data Entry), User Training (report generation and analysis) at Brampton and Calgary facilities.

EDUCATION & CREDENTIALS

Project Management Professional (PMP), Project Management Institute	2002
Bachelor of Arts, York University, Toronto, Ontario	1995

Major: Industrial-Organizational Psychology
Minor: Business Administration, Research

Electronics Engineering Technologist, Humber College, Toronto, Ontario	1983

Major: Telecommunications
Minor: Programming, Statistics
Further Details on Management Training, and Technical Proficiencies & Experience available on request.

CORPORATE LEVEL PERFORMANCE AWARDS

Process Development Award, Teamwork	1997
ABC Networks President's Awards of Merit: Excellence	1995
ABC Networks President's Awards of Merit: Teamwork	1992
Value Engineering, Cost Improvement	1991
ABC Networks President's Awards of Merit: Administrative Excellence	1987

Senior Engineer After

John Walton
EET, BA, PMP

Picture of the
award
xxxxx
President's Award
of Merit:
Excellence 1995

Picture of the
award
^^^^^
President's Award
of Merit:
Teamwork 1992

Picture of the
award
xxxxx
President's Award
of Merit:
Administrative
Excellence
1987

OBJECTIVE
To increase shareholder value by developing and delivering high-quality products.

PROFILE
Fifteen years' experience in both the technical and business aspects of the technology industry, with a demonstrated ability to deliver.

How:
- Discover What the Customers Want.
- Drive Design, Development, and Delivery.
- Act as Technical Evangelist Where Necessary.
- "Six Sigma" Methodology.

Examples:
➢ Created XXX's first OC3/OC12 FPGA that was functional across multiple SONET-Ethernet interface 'Madonna Metro' products.
➢ Improved XXX's BMS100-Family product line integrity and robustness by 60%; achieved a cost improvement of $30M.

CAREER DRIVER
Inspiring and leading teams to develop breakthrough products, which solve customer demands and have real commercial value in the global market.

SPECIAL SKILLS
My experience has honed the following development know-how:
- Execution—regularly delivering to fixed time schedules against all odds.
- Experimentation—relentless probing for new R&D and product approaches.
- Expressive clarity—strategic development plans.
- Management—optimizing people and finances to meet objectives.
- Strategic Alliances—technical liaison throughout project life cycles.
- Budget maximization—for effective use of finite resources.
- Leadership—of teams ranging from 6-30 people across multi-country-sites.

EMPLOYMENT HISTORY
XXXXXXXXX   **2001–present**
Private consulting focused on bringing technology to market *"faster, better, cheaper."*

ABC Networks Limited   **1983–2001**
FPGA/Hardware Designer - Maddonna Packet Switching, xxxx, 1999–2001
Sr. Projects Leader - Technology (R&D) Organization, zzzz, 1999–94
Projects Leader - Technology (R&D) Organization, zzzzz, 1993–90
Task Force Leader - Quality—Customer Satisfaction, zzzzz, 1989–87
Analyst - Quality—Reliability Engineering, zzzzz, 1986–84
Functional Tester - Customer Service Operations, zzzzzz, 1983

EDUCATION
Project Management Professional (PMP), Project Management Institute, 2002
Bachelor of Arts, XXXX University, XXXX, Ontario, 1995
Major: Industrial-Organizational Psychology
Electronics Engineering Technologist, XXXXXXX College, XXXXXX, XX, 1983
Major: Telecommunications

——— Appendix F: eXtreme Resume Makeovers ———

Getting an employer's attention is the focus of this section.

On the following pages, you'll find several resume makeovers that have been used by real people to land interviews and, ultimately, job offers. They range from president to programmer, from VP of sales to volunteer coordinator.

Please note: All of the resumes in this section are intended to run on one page. However, because of space constraints, the following sample flows onto a second page here.

◆◆◆

MARK
SMITH

LOGO

*Co-honored as #1
outstanding
employee within
company of 110 staff*

*Consistent achiever
of 110% productivity
awards*

*Developed and
delivered a sales
campaign that
created 23%
response rate from
top Canadian CEOs
in tough economic
conditions.*

*Developed and
Managed a #1
highest volume store
for a 270-store chain
between $4 and $5
million in revenue
per year.*

PHONE: xxxxxxxx
CELL: xxxxxxxxx
FAX: xxxxx
E: xxxx

xxxxx
ANY TOWN

En*tel*e*chy – becoming actual what was only potential.

SUMMARY

Dynamic sales operations manager. Accomplishments include:
- ➤ Integral part in making company best investment in Canada rated by Finance Forum.
- ➤ Key initial contact and presenter to help win $22 million US in venture investment.
- ➤ Merged multi-million dollar CAN., UK, and US sales operations into 1 sales infrastructure.
- ➤ Produced the case studies and business proposals for many large contracts.
- ➤ Developed collateral that was integral to all corporate contracts.
- ➤ Key initial presenter with majority of business partners (i.e. Novell).
- ➤ Consistently created high response rate sales campaigns and lead generation programs.

CAREER DRIVER

Taking the surety of success, the passion to succeed, and the deft handling of economic drivers to build great organizations.

SPECIAL SKILLS

My start-up and management experience has honed the following skills:
- ➤ Execution—consistently delivering to fixed time schedules against very steep odds.
- ➤ Creation—business plans, case studies, marketing documentation, reports, collateral, and presentations which always enjoy extensive internal/external usage.
- ➤ Service Excellence—in creating new clients and maintaining established relationships.
- ➤ Focus—efforts always stay targeted on what creates income and reduces costs.
- ➤ Management—finding & hiring people with the highest potential & then actualizing it.
- ➤ Achievement—delivering performance in the top 1% in all previous positions.
- ➤ Strategic Budgeting—creating best usage of finite resources.
- ➤ Communication—effective trainer in sales procedures, software, and product services.
- ➤ Administered—SFA (Sales Force Automation) systems.

EMPLOYMENT HISTORY

Consultant CRM solutions—ABC Technologies (2002–2003)

Sales Operations Manager—DEF Networks Ottawa ON, (1998–2002)
- ➤ A leader in online interaction software and Internet phone technology.
   - ◆ sales ◆ reporting ◆ collateral development ◆ customer service
   - ◆ marketing ◆ training ◆ procedure development ◆ strategy

Manager—Consumers Distributing (1984–1998)

EDUCATION
Bachelor's degree in Economics
Diploma in Business Administration
Wilfrid Laurier University, Waterloo, Ontario

Microsoft Certified Professional
Microsoft Certified Trainer

PRODUCT EXPERTISE
* MS Excel   * MS Word   * MS PowerPoint   * Goldmine 5.7
* Working knowledge MS Project, Access, Visio, and Adobe Products
* Upshot

MARK SMITH

logos

cat·a·lyst - an agent that provokes or speeds significant change or action.

SUMMARY

Dynamic senior executive. Accomplishments include:
- ➢ Ignited turn-around of mid-sized high-tech company from insolvency to profitable double-digit growth and positive cash flow in 12 months.
- ➢ Drove a pipeline increase in OEM-business from US $8M – $20M in 24 months.
- ➢ Developed vision and bus plan to close mission critical technology license agreement.
- ➢ Merged US $40M operations without any loss of business and staff.
- ➢ Defined and launched distribution partners and strategic alliances in 20+ countries.
- ➢ Positioned international subsidiaries for North American corporations resulting in triple digit international growth numbers.

CAREER DRIVER

Introducing new technologies into markets and positioning a corporation as the market leader.

SPECIAL SKILLS

My international sales and start-up experience has honed the following skills:
- ➢ Execution—regularly delivering to fixed time schedules against all odds.
- ➢ Experimentation—relentless probing for new R&D and product approaches.
- ➢ Expressive clarity—industry commentary, outreach, Web writing.
- ➢ Management—optimizing people and finances to meet objectives.
- ➢ Strategic Alliances—to bring products to market faster.
- ➢ Budget maximization—for effective use of finite resources.
- ➢ Leadership—of teams ranging from 8–90 people

EMPLOYMENT HISTORY

Contract Executive xxxxxxxxxxxxxxxxxxxx **(1998–2002)**
- ➢ The leading supplier of Micro-laser Manipulation systems for biomedical applications.

♦strategy	♦sales	♦ business development
♦marketing	♦OEMs	♦ contract negotiations

President & CEO—xxxxxxxxx (1992–1998)
Managing Director—xxxxxxxxxxx (1990–1992)
General Manager—xxxxxxxxxxx (1987–1990)
Director Marketing Services—xxxxxxxxxxxxx (–1987)

EDUCATION

Bachelor's degree in Computer Science and Electronic Engineering
Engineering College xxxxxxxx, Germany

Graduated as Professional Engineer
xxxxxxxxxx, Germany

BOARD OF DIRECTORS
- ➢ Cofounder of xxxxxx Association, technology association for xxxxxxx
- ➢ Director of the Board and Chairman of the Marketing Committee of xxxxxxx

PHONE: xxxxxxxx

E: xxxx

xxxxx
xxxxxxxxxx

Mark Smith

"Mark, in all my dealings with you I have always been impressed with your ability to distil information down into actionable items and then just do it."

David

Summary

Complex sales and business development professional. Accomplishments include:

- Opening $300M+ in net new opportunities for account team.
- Closing $16M in new carrier business.
- Developing marketing material, cost justification analyses, and product direction.
- Accelerating public awareness and brand marketing.
- Igniting new product sales 6 weeks after joining.
- Managing development of Web marketing and e-business platform.

Career Driver

I seek a position that will allow me to promote and sell the promise of developments in the new age of high technology to business targets and leading opinion-makers, for an organization that has an aggressive mind-set.

Special Skills

Working with Synamics, Nortel, and RIM, has honed these skills:

Sales—track record of leading teams to successful outcomes.

- Conceptualization—preparing and presenting deals at the CEO/CFO/CIO level.
- Creative concepts—branding, e-mail marketing, partnerships.
- Experimentation—relentless probing for new revenue approaches.
- Budget maximization—for effective use of finite resources.
- Management—optimizing people and finances to meet objectives.

Employment History

Business Development Manager **(2001–2002)**

xxxxxxxx - Toronto

◆sales	◆product direction	◆marketing
◆analysis	◆delivery	◆business development

Business Development Manager, xxxxxxxx networks (1999–2001)
Independent Consultant (1998–1999)
- Clients included: xxxxxxxx, xxxxxxxx, xxxxxxxx, and xxxxxxxx.
Marketing Specialist/Product Manager, xxxxxxxx (1997 contract)

Education

xxxxxxxxxxxx University, 2003, part-time studies toward B.Sc.
xxxxxxxxxxxx University, 1997, Bachelor of Economics

Professional Development

xxxxxxxxxxx Leadership—Leadership and management mentoring and training
Target Account Selling—Professional sales account planning
Target Opportunity Selling—Professional sales individual opportunity planning and execution
Sales Presentations @ xxxxx Networks—Presentation planning and delivery to open and close large, complex opportunities
New Strategic Selling—Account, territory, and opportunity planning and execution for complex sales

xxxxxxxxxxxx
xxxxxxxxxx

MARK SMITH

LOGO

LOGO

LOGO

LOGO

SUMMARY

Over 20 years of experience as a leader delivering on start-up, turnaround, and growth challenges in a myriad of markets and economic conditions. Demonstrated ability to combine sales and marketing issues effectively, transforming complex technology products and services into vibrant, compelling, and easily understood high value business propositions.

Executive accomplishments include:
- Developed a *Global Strike Team* to rapidly engage customers in the FP1000.
- Delivered **triple digit growth** numbers.
- Ignited sales for a US multinational, closing **$6 million** year one.
- Drove a pipeline increase in new business from **$1.7 to $9 million**, year one.
- Advised strategic team to close mission critical **$9 million** VC funding.
- Negotiated multiple million-dollar agreements with **Tier 1 Strategic Partners.**

SPECIAL SKILLS

International sales and start-up experience have honed the following skills:
- Execution—regularly delivering to fixed time schedules against all odds.
- Experimentation—relentless probing for new sales/partnership approaches.
- Management—optimizing people and finances to meet objectives.
- Creative concepts—branding, e-mail marketing, partnerships (VAR/OEM), road shows, strategic partner seminars.
- Strategic Alliances—to capture marketshare faster.
- Budget maximization—for effective use of finite resources.
- Leadership—of teams ranging from 8 to 20 people.

SELECT ENTERPRISE CLIENTS SOLD

Client, Client

EMPLOYMENT HISTORY

2000 to 2002	General Manager, xxxxx.
1998 to 2000	General Manager, xxxxxxxx
1996 to 1998	Branch Manager and Canadian Business Development Manger, Public Sector & Health Care, xxxxxxxxxx
1994 to 1996	Regional Sales Manager, xxxxxxxxxxx
1993 to 1994	Director Sales and Marketing, xxxxxxxxxxx
1991 to 1993	Managing Director, xxxxxxxxxx.
1987 to 1991	Area Manager, xxxxxxxxxxxxxxxxx.
1984 to 1987	Account Executive/Product Specialist, xxxxxxxxxxxxxx.
1982 to 1984	xxxxxxxxxxxxxx Account Manager

Mark Smith
M.Sc.Eng.

Objective

To increase shareholder value by driving sales results in xxxxxxx and xxxxxx.

Summary

Dynamic senior sales and marketing executive. Milestones include:
- ➤ Developed a *Global Strike Team* to market and sell in Eastern Europe.
- ➤ Led negotiations and signed contracts with the 3 dominant national pharmaceutical and cosmetics plants as well as export companies to provide pharmaceutical raw materials.
- ➤ Grew start-up sales distribution company from $x to $x in x years.
- ➤ Developed a regional distribution network for pharmaceutical and cosmetic products.
- ➤ Initiated and leveraged worldwide partnerships to expand markets in Europe and US.

Special Skills

Working within the private and public sectors at the local, national, and international levels has honed these skills:
- ◆ Sales—track record of leading teams to successful outcomes.
- ◆ Conceptualization—preparing & presenting deals at both the "C" and End-user levels.
- ◆ Experimentation—relentless probing for new revenue approaches.
- ◆ Budget maximization—for effective use of finite resources.
- ◆ Management—optimizing people and finances to meet objectives.
- ◆ Project Management—evaluate programs, initiate projects, execute strategies.
- ◆ Creative—sales strategy including: value proposition, positioning, collateral.

Employment History

Canada:

Software Designer, xxxxxxxxxxxx, zzzzz	12/01–11/02
Software Designer, xxxxxxxxx, zzzzzz	02/01–11/01

Europe:

President - xxxxxxxx, zzzzzzzzzzzzz 12/96–12/00
- ➤ *Marketing Advisor* - xxxxxxxxxxxxxxx 04/00-07/00
- ➤ *Software and Sales Engineer* - xxxxxxxxxx, zzzzzzzzzzz 01/97-04/00
- ➤ *Regional Sales Representative* - xxxxxxxxxx, zzzzzzzzzzzzz 03/96-12/96

Regional Sales Representative - xxxxxxxxxx, zzzzzzzzzzz 04/95-02/96
Trade and Marketing Manager - xxxxxxxxx, zzzzzzzzzzzzz 11/92-04/95
Project and Sales Engineer - xxxxxxxxxx, zzzzzzzzzzzzz 02/92-11/92
Software Designer, xxxxxxxxxxxxxxx, 11/90-02/92
Associate Assistant, Technical University, xxxxxxxxxxxxx, 09/89-06/90
Software Designer, xxxxxxxx Academy – Automation, xxxxxxxxx 11/88-11/90

Education

M.Sc., Technical University – xxxxxxxxx, Engineer in Biotechnology 84-89
    (Comparative Education Service Certified, Univ. xxxxxxxx – CES)
Certificate - International Economic Relations, Technical Univxxxxxxxxxxxxx, 87-89

Personal

Language Skills: Fluent English, Bulgarian, and Russian (oral and written).
Professional Associations: xxxxx Academy of Sciences, xxx Association.
Security Clearance: Cleared to Enhanced Security Level; Secret Security Level pending.

xxxxxxxxxx
xxxxxxxxxxxx
xxxxxxxxxx

Mark Smith

Cell: xxxxxx
Office: xxxxxx
Home: xxxxxxxxxx
http://www.xxxxxx
abc@def.com

rain·mak·er – one whose influence can initiate progress or ensure success.

SUMMARY

Technology sales and busdev executive. Milestones include:
- Sold a two-year global software contract to xxxxxxxxxx valued at $10,000,000+.
- Developed a *Global Strike Team* to rapidly engage senior level management at numerous Level 1 financial institutions including Client, Client, Client, Client, Client, Client, Client, Client, Client, Client, Client.
- Initiated leveraged worldwide partnership/ relationships with Client (London), Client (Geneva), Client (Hamburg), and Client (Ottawa).
- Licensed 80,000+ copies of Product to Client Computing.
- Closed multiple million-dollar agreements with OEM's: Client, Client and Client.
- Closed unique Software Licensing Agreement with Client centered on a transactional pricing model for it's customer, Client.
- Invented the product for the Productä. Generated orders for over 10,000 units during the first month.

CAREER DRIVER

An opportunity that will allow me to promote the promise of developments in the new age of high technology to business for an organization that has an aggressive mind-set.

EMPLOYMENT HISTORY

TWA, President (2001–present)

➢ Private consultancy with a mandate to push the envelope for business development and leverage market entry in the fast-changing high-tech world.

| ◆strategy | ◆sales | ◆business development |
| ◆marketing | ◆OEMs | ◆contract negotiations |

Vice President Strategic Business Developmen— Company (1999–2001)
Founder &Vice President Sales & Marketing—Company (1997–1999)
Director Global Business Development—Company (1992 –1997)
Senior Account Executive—Company (1989–1992)
Manager of National Accounts, Financial Services—Company (1986-1989)

EDUCATION

Nice College

Mark Smith

Objective

To double or **triple the number of new sales leads and customers you sell** each month—without increasing your marketing budget.

Over 30 years' experience in both technical and business aspects of the technology industry, with a demonstrated ability to Light **A Fire Under Customers to Make Your Sales Copy Sell**... *even In a Recession.*

How:
* Discover What Your Customers Want and You Offer;
* Write Motivational Copy That Sells;
* Create Sales Letters That Make Your Customers <u>Want Your Product</u>... then Buy!

Career Driver

Inspiring prospects with powerful persuasive words to get a move on and purchase now—before they get sidetracked.

Special Skills

Experience has honed copywriting know-how:
* Creative concepts—branding, e-mail marketing, partnerships.
* Experimentation—relentless probing for new revenue approaches.
* Expressive clarity—industry commentary, outreach, web writing.
* Management—optimizing people and finances to meet objectives.
* Budget maximization—for effective use of finite resources.

Employment History

Self-employed (05/97–present)

·strategy   ·business development   ·marketing

Company City State, Computer Specialist (85–97).
Company City State, Customer Engineer (80-84).
Company City State., Senior Programmer (04/80-08/80).
Company City State, Staff Accountant (78-80).
Company City State business systems architect (74-80).
Company City State, Project Engineer (01/73–11/73).
Company City State, Engineer (71–73).
Company City State, Physicist (68–71).

Education

BA Physics, Nice State University
Management Engineering College
* Business Process Re-engineering
* Functional Process Improvement

Call Mark Smith @ xxxxxxxx

Right now, if time is at a premium and you're in a crunch period, if there simply aren't enough hours in the day for you to do the job yourself, contact me.

Mark Smith

LOGO

Corporate priorities dwell at the intersection of private interests and public concerns—where issues and interests collide, and where leaders meet to frame government policy. You can only affect the telecommunications debate if you make your message clear, make your voice heard, and make every opportunity count.

SUMMARY

Twenty years' experience as a lawyer and policy advisor with a recognized ability to influence key government and private sector stakeholders. My most significant successes include:

- Federal government's point man for long distance competition hearings.
- Policy advisor for the creation and execution of the Broadcasting Act.
- Spearheaded the ABCD-TEL XXXXXXXX Study on Internet traffic.
- Positioned and wrote the XXX Declaration for the reform of XXXX.
- Influenced Group to adopt zzzzzzzzzz.

SPECIAL SKILLS

My Government Relations experience has honed these skills:

- Clarity of expression—industry commentary, outreach, legislation, submissions.
- Conceptualization—analyzing structures and policies.
- Leadership—optimizing people and stakeholders to meet objectives.
- Execution—regularly delivering to fixed time schedules and during crises.
- Negotiations—national telecommunication issues inside and outside government.

EMPLOYMENT HISTORY

Company of Consultants, **Principal (1989–present)**

- Regulatory proceedings before the telecommunications regulators and policy makers;
- North American numbering plan activities;
- Spectrum allocation;
- The zzzzzz Act;
- The xxxxxxxxxxx Act;
- The qqqqqqqqqqqqq Act.

Policy Advisor, Company	(87–89)
Senior Associate, Company	(80–86)
Assistant to the Vice-Chairman, Telecommunications, Company	(78–80)
Staff Writer, xxx Commission on xxxxxx and xxx	(76–78)
Machinery of government directorate, xxxxxxxxxx	(74–76)

EDUCATION

Bachelor of Civil Law, Nice University, 1973
Bachelor of Arts (Honours) in political science and history, 1970

BOARDS

Some Internet Registration Authority 2001–2006

ASSOCIATIONS

Bar of Great City, 1974–1997.
Nice Society of xxx, president 1993–1995.

LANGUAGES

French and English, spoken and written.

Mark Smith

xxx-xxx-xxxx

LOGO

Client

logos

down

this

section

Xxxxxxx
Xxxxxxxxxx
Xx

OBJECTIVE

To increase shareholder value by developing high quality commercial grade optical products.

PROFILE

Optoelectronical engineer with five years experience in optical components for DWDM/CWDM network applications, measurements, test techniques, and process automation.

ACCOMPLISHMENTS

* Developed new tooling method (process and software), which increased yield by 50%, saving ABC $12.5M/year.
* Redesigned and improved Filter Tuning process, which increased Yield from 45% to 85%, saving ABC $10M/year.
* Designed full NPI process flow with checkpoints and approvals, which eliminated catastrophic failures, saving on average $10K/week.

CAREER DRIVER

Inspiring and leading teams to develop breakthrough products, which solve customer demands and have real commercial value in the global market.

SPECIAL SKILLS

* Execution—regularly delivering to fixed time schedules against all odds.
* Experimentation—relentless probing for new R&D and product approaches.
* Management—optimizing people and finances to meet objectives with customers.
* Strategic Alliances—to bring products to market faster.
* Budget maximization—for effective use of finite resources.
* Leadership—of teams ranging from 3–30 people.

EMPLOYMENT HISTORY

ABC Company, Any Town   Jan. 99–Apr. 03
* Joined as a product engineer in the filter group and was quickly promoted to team leader after demonstrating ability to manage a group of 5 engineers. Remained the key application engineer representing the OCLI Thin Film Filters product group after successfully transitioning the department to California.

Sr. Application Engineer   Aug. 01–Apr. 03
Team Leader – Project Management   Sept. 99–Aug. 01
Product / Process engineer   Jan. 99–Sept. 99

SALON DE PROVENCE AIR FORCE BASE, Salon, France   Sept. 97–Jun. 98
* Completed 1-year mandatory army service, specialized in IS and communications.

xxxx, Montpellier, France   Feb. 97–Jun. 97
* Completed a 5-month work placement program, designing and implementing a laser telemeter—a fusion of optical, electronic, and mechanical engineering.

EUROPEAN CENTER OF xxxxxxxxxx   Apr. 94–Jun. 94
* Completed a 10-week work placement program, designing and implementing a circuit board based on Phase Lock Loop technology.

EDUCATION

Masters of Sciences in Optoelectronics   Sept. 94–Jun. 97
* Nice University of Sciences, Montpellier, France

Bachelor of Science in Electronics   Sept. 92–Jun. 94
* Nice University of Sciences, Grenoble, France
* University Degree of Technology, specialization: electronics & computer engineering.

Susan Star

cat·a·lyst - an agent that provokes or speeds significant change or action.

SUMMARY

Over 20 years' experience as a leader in both technical and business aspects of the technology industry, with a demonstrated ability to manage and motivate people, create commercially successful products, and develop and execute strategies that result in marketplace success. My most recent successes include:

- Developed vision and bus plan that brought VCs to the table (due diligence continues).
- Executed new technology roadmap and delivered product to market on time.
- Drove a pipeline increase from $100,000 to $7 million within 8 months.
- Defined and launched two new products into a new vertical.
- Designed and executed client acquisition plan resulting in 45% increase in revenue.
- Ignited new product revenue 400%.
- Positioned company to win the "Crossroads – Best product in category" award.

CAREER DRIVER

President/CEO of an emerging Internet company that has an aggressive mind-set.

SPECIAL SKILLS

My start-up experience has honed these skills:
- Management—optimizing people and finances to meet objectives.
- Budget maximization—for effective use of finite resources.
- Creative concepts—branding, e-mail marketing, partnerships.
- Experimentation—relentless probing for new revenue approaches.
- Expressive clarity—industry commentary, outreach, Web writing.
- Conceptualization—analyzing structures and policies.

EMPLOYMENT HISTORY

Company, President (08/01–present)

◆strategy	◆communications	◆investor strategy
◆marketing	◆sales	◆business development

Company, President – Ottawa, ON (10/00–08/01)
Company, VP Marketing – Ottawa, ON (11/97–10/00)
Company, VP Marketing – Ottawa, ON (01/97–11/97)
Company, Director Data and Voice Services (95–97)
Company, Marketing Director (92–95)
Company, Founding Partner (89–92)
Company, Marketing Services Manager (88–90)

EDUCATION

Ph.D.   University of Toronto—1987
M.Phil.   University of Waterloo—1982, M.A.—1980, Hon. B.A.—1978

BOARD OF DIRECTORS

➢ Software Human Resource Council
➢ xxxx
➢ xxxxxxx Public Relations

Xxxxxxx
Xxxxxxxxxx
Xx

Mark Smith

LOGO

Quote from

the press

OBJECTIVE

To increase shareholder value by developing high quality commercial grade software products.

PROFILE

Over 20 years' experience in both technical and business aspects of the technology industry, with a demonstrated ability to develop new software products (9 to date) for both commercial and military markets.

How:
- Discover What the Customers Want;
- Drive Design, Development and Delivery;
- Act as Technical Evangelist Where Necessary.

CAREER DRIVER

Inspiring and leading teams to develop breakthrough products, which solve customer demands and have real commercial value in the global market.

> "Mark is a visionary. He is able to create a vision for a product and deliver upon that vision." (D.D.)

SPECIAL SKILLS

Experience has honed development know-how:
- Execution—regularly delivering to fixed time schedules against all odds.
- Experimentation—relentless probing for new R&D and product approaches.
- Expressive clarity—strategic development plans.
- Management—optimizing people and finances to meet objectives.
- Strategic Alliances—to bring products to market faster.
- Budget maximization—for effective use of finite resources.
- Leadership—of teams ranging from 8 to 55 people.

EMPLOYMENT HISTORY

Company, **Senior Manager of Development, (97–present)**

- Recruited to turnaround the development of a legacy product line. Launched new technology road map and subsequent product, which catapulted **Company** to the front of the line in a new category.

Company, Vice President of Research and Development – New York, NY (96–97)
Company, Supervisor Of Technical Support – New York, NY (94–96)
Company, General Manager/Product Manager, Manager Engineering, Engineer– New York, NY (89–94)
Company, Research Associate – New York, NY (87–89)
Company, Software Engineer – New York, NY (86–87)
Company, Manager – New York, NY (85–86)

EDUCATION

Bachelor of Computer and Systems Engineering
Nice University, 1985

MARCY SMITH

JOB OBJECTIVE

Casual or Substitute teacher, on-call for either French or English K–6 children.

PERSONAL SUMMARY

Creative, energetic, resourceful, mother of two school-age children.

SPECIAL SKILLS

My courses in **Early Childhood Education** and **Psychology** with my work experience in the service industry have honed these skills:

- Training—delivered new employee indoctrination.
- Management—influence and optimize people to meet objectives.
- Budget maximization—for effective use of finite resources.
- Experimentation—relentless probing for novel solutions and approaches.
- Organization—canvass volunteers, build and motivate teams, direct outcomes.
- Project management—evaluate programs, initiate projects, develop and execute strategies.
- Computer—Windows, Internet, Microsoft Office.

EDUCATION

University of Ottawa—part-time, 2 years of Psychology
Heritage College—Early Childhood Education, interrupted by maternity leave
CEGEP de l'Outaouais—Science Sociale December 1993

EMPLOYMENT HISTORY

04/97–	*Maternity Leave*
09/95–04/97	**Company**, Senior Customer Service Representative Beneficila specializes in high-risk loans to middle income families. I was "Employee of the Month" several times and as such was responsible for training new employees.
05/91–04/95	**Company**, Bartender and Waitress
10/90–05/91	**Company**, Waitress
89–90	**Company**, Receptionist

LANGUAGES

Fluent in French and English

COMMUNITY INVOLVEMENT

President of Neighborhood Watch Association

Mary Moore

LOGO

Objective

To increase shareholder value through correct accounting practices.

Summary

Creative, energetic, experienced accounts-payable professional.

Special Skills

Working in the accounts payable group at Future Electronics has honed these skills:
- Audit—keen sense for catching discrepancies.
- Organization—extensive expense audit and month-end reporting.
- Budget maximization—for effective use of finite resources.
- Organization—canvass volunteers build and motivate teams, direct outcomes.
- Customer service—reduced claim errors and improved reimbursement times.
- Management—influence and optimize people to meet objectives.

Employment History

Accounts Payable Officer—Company, **City, State**　　　　　　　**(1996–10/020)**
> Audited and processed expense claims for 1100 salespeople and senior management across North America. Acted as department supervisor to cover a maternity leave. Extensive background in expense auditing and the preparation of detailed month-end reports. Annual revenues ~$9.6 billion.

Receptionist—**Company** (08/95–02/96)
Receptionist/Clerk—**Company** (10/94–08/95)

Computer Skills and Formal Education

Excellent knowledge of Microsoft Office and Oracle Financials. Proficient in various packaged software applications, including word processing, e-mail, etc.

Nice Institute, Nice City - Office Systems - A.E.C.
Nice High School, Nice City - Diploma

Languages

Fluent in English and French

Interests and Hobbies

Reading, music, interior design, and travel

Xxxxxxxxx
Xxxx
x
xxxxxxxxxx

Mark
Smith

xxx-xxx-xxxx

LOGO

Summary

High-energy, results-focused coax cable/fibre optic technician with 18 years experience.

Accomplishments

➢ Developed Triolithic Return Alignment Technique.
➢ Responsible for rapid analyses and repair for critical users like: Client Headquarters, Client, ATM banking system, Internet, and traffic lights.
➢ Received Company *Service Award* for outstanding work on the ZZZ Rebuild.

Special Skills

Working in a fast-paced service-oriented environment has honed these skills:

➢ Technical audit—keen sense for system-wide troubleshooting.
➢ Execution—exceptional track record for repair times - 25% faster than company average.
➢ Project management—working with state-of-the-art high impact systems.
➢ Customer relations—focal contact for commercial customers during critical system failures.
➢ Training—inspired team commitment to quality service and customer satisfaction.
➢ Organization—meticulous record keeping and reporting.
➢ Budget maximization—for effective use of finite resources.
➢ Management—influence and optimize people to meet objectives.

Employment History

Company xxx/company zzz–*City, State*   *1985–2002*

➢ Joined as a maintenance technician troubleshooting cable lines and was promoted quickly through a series of increasingly demanding technical jobs including quality control and duty supervisor to become 1 of only 4 advanced network technicians who maintain Rogers' large and extremely complex network system throughout zzz and zzzzz.

Network Maintenance Technician	*1995–2002*
Quality Control Technician / Duty Supervisor	*1990–1994*
Maintenance & Senior Maintenance Technician	*1985–1990*

Education and Additional Training

Excellent knowledge of Microsoft Office and proficient in related software applications.

➢ Electronics Engineering Technician Diploma—Nice College, 1982–1984
➢ Electronics Engineering Technologist Program—Nice College, 1979–1981
➢ Fibre training and online courses on Company network 1995–2002
➢ Linear Circuits, Nice College, 1990
➢ Basic CATV Concepts, Nice College, 1988

Professional Certification

➢ **Security Clearance**
➢ Manhole Access License
➢ Skyvan Hydraulic Aerial Device License

Xxxx
Xxxxxxxxx
xx
xx

Technical Skill Summary

Emergency cable locates; trunk sweeps; cable/fibre equipment transmitters/receivers and power supplies; amplifiers, modulators, spectrum analyzers, video surveillance cameras; use of OTDR, laptop computer, and construction plans to pinpoint cable/fibre failures; gas detectors, water pumps, and other manhole equipment; system monitoring and manual checks including system channel signals, C.T.B., cross-modulation, carrier to noise, and hum measurements; and Quality control.

Mark Smith

LOGO

SUMMARY

Fourteen years' experience as a leader in both technical and business aspects of product design, with a demonstrated ability to manage and motivate people, create commercially successful products, and develop and execute strategies which result in marketplace success.

My most recent successes include:
- Developed vision and business plan to move 80,000–100,000 units per item.
- Designed and launched new products into three new verticals.
- Drove distribution across North America.
  - Direct to small independent retailers and larger retail chains like Pet Value; OEMs to retailers like Client, Client; currently negotiating with the Client for their XYZ stores.

SPECIAL SKILLS

My experience in bringing new products to market, from design and development, to prototyping, and offshore manufacturing has honed these skills:
1. Leadership—evaluate programs, initiate projects, develop and execute strategies.
2. Management—influence and optimize people to meet objectives.
3. Research & Development—consultant (gun-for-hire) on commercial products.
4. Budget maximization—for effective use of finite resources.
5. Experimentation—relentless probing for novel solutions and approaches.

EMPLOYMENT HISTORY

1997–   *President/Industrial Designer*—Company, City.
➤ Focus on product design, marketing, and consulting.

1989–1997   *Director of Design and Marketing*—Company Ventures and Innovations Group
➤ Responsible for the assembly team, equipment, parts resource all design, manufacturing, all research and development.

1977–1989   *Fire Fighter*—City and State
➤ Fire fighter, and emergency rescue, cars, suicide, explosions. Youngest member to join. Retired first-class rank.

EDUCATION

Bachelor of Science, Nice University, 1986

PATENTS

5083108 – Bike Horn—a large volume product, over 100,000 units sold so far.
5535804 – Pet Door—volume is about 90,000 units sold so far.
5490346 – Fishing Lure—just released.
➤ Generally speaking the capital generated from these units ranges from $4 to $6 net, depending on the method of distribution.

COMMUNITY INVOLVEMENT

Scouts—Aylmer, Quebec Beavers & Cub Scouts 1999–2002
ZZZZ Hospital, Juvenile Offenders Program, 1980–1983
Big Brothers, 1984–1986

Xxxxxxx
xxxxxxxxxx
Xxxx

Mary Moore

Summary

Creative, energetic, experienced community builder and public school educator.

Special Skills

Working within the education system and various community organizations at the municipal, provincial, and federal levels has honed these skills:

- Organization—canvass volunteers build and motivate teams, direct outcomes.
- Project management—evaluate programs, initiate projects, develop and execute strategies.
- Leadership—community outreach, public speaking, consensus building.
- Management—influence and optimize people to meet objectives.
- Budget maximization—for effective use of finite resources.
- Teaching—assess students, develop curriculum, facilitate learning, test, and reassess.
- Experimentation—relentless probing for novel solutions and approaches.

Employment History

Teacher—ABC School Board, City/State (1995–present)
> Have taught: Junior Kindergarten, Kindergarten, Grade 1-2 split, Grade 2 and Grade 6 students across the Pontiac Valley. Physical Education Instructor at Eardley Elementary for one year.

Director —Company, self-directed Preschool, Aylmer, Que. (1993–1995)
Teacher—Company Board of Education Adult Education & E.S.L., Ottawa, ON (1993–1996)
Enumerator & Returning Officer—Company. (1989–1993)
Executive Assistant to Minister of Education—Company Regina, Sask. (1984–1985)
Policy Advisor—Premier of Saskatchewan. (1983–1984)
Director French Kindergarten—Company, Baie d'Urfe, Que. (1982–1983)
Lab Technician—Company, Lachine, Que. (1980–1982)
Supply Teacher—Company, Beaconsfield, Que. (1980)

Education

University of Calgary, B.Ed. Early Childhood Education & Educational Psychology (1975–1980)
Mariannopolis College, C.E.G.P. Diploma – Languages and Psychology (1972–1974)
Additional post-graduate courses in Psychology & E.S.L. (U-Sask., Carleton, McGill) (1984, 91, 98)

Languages

Fluent in English and French

Community Involvement

Ottawa Twin Parent Association, Director/Counselor—for higher order of multiples, (1990–2000)
Aylmer Women's Forum, Coordinator—"Speakers Bureau" and community outreach (1986–1990)
South Hull School Committee, Chair—speakers, information meetings (1997–2000)
Eardley School Committee, Chair—develop and maintain daily volunteer base (1993–1996)
Eardley School Committee, Parent Committee Representative (1997–1998)
Literacy Council, Member/Volunteer—supporter and canvasser, (1997–present)
Medical Council of Canada, Volunteer (1998–present)
Ottawa Board of Education, Volunteer Teacher (1990–1991)

Xxxxxxx
xxxxxxx
xxxxxxxxxxx

Mark Smith

cat·a·lyst - an agent that provokes or speeds significant change or action.

SUMMARY

Over 25 years' experience as a leader facing start-up, turnaround, and growth challenges in a myriad of markets and economic conditions. Demonstrated ability to combine engineering and marketing backgrounds effectively, transforming sterile products and services into vibrant, compelling, and easily understood, high-value propositions.

My most recent successes:
- Ignited business units at Company, generating $130M in incremental revenue.
- Built the Entelechy Group from concept to over $35M in revenue.
- Drove Company from insolvency to $11 million and acquisition.
- Led 22 product development teams at Company, generating $145M in incremental sales.

CAREER DRIVER

President/CEO of an emerging company with an aggressive mind-set.

SPECIAL SKILLS

Successful ventures in entrepreneurial and corporate settings have honed these skills:
- ROI/Shareholder equity—have built companies using only retained earnings.
- Management—optimizing people and finances to meet objectives.
- Budget maximization—for effective use of finite resources.
- Creative concepts—branding, e-mail marketing, partnerships (VAR/OEM).
- Experimentation—relentless probing for new revenue approaches.
- Expressive clarity—boardroom, industry commentary, outreach.
- Conceptualization—analyzing structures and policies.

EMPLOYMENT HISTORY

Company	General Manager, Joint Venture Business	1998–2001

- strategy
- marketing
- communications
- sales
- investor strategy
- business development

Company	President & Chief Executive Officer	1988–1997
Company	Managing Partner	1982–1988
Company	Director of Marketing	1981–1982
	• Marketing Manager	1979–1981
	• Product Planner	1978–1979
Company	Powder Plant Manager	1976–1978
Company	Ethylene Plant Area Engineer	1974–1976
Company	Halogens Production Support Engineer	1973–1974

EDUCATION

MBA	Marketing	Michigan State University	1978
BS	Chemical Engineering	Purdue University	1973
USA	Marine Corps	2nd Marine Force Reconnaissance	1970–1971

address
address

BOARD OF DIRECTORS

➢ xxxxx

Mark Smith
EET, B.A., PMP

Picture of the award
xxxxx
President's Award of Merit:
Excellence 1995

Picture of the award
^^^^^
President's Award of Merit:
Teamwork 1992

Picture of the award
xxxxx
President's Award of Merit:
Administrative Excellence
1987

xxxxxxxxx
xxxxxxx
xxxxxxxxxx

OBJECTIVE

To increase shareholder value by developing and delivering high quality products.

PROFILE

Fifteen years' experience in both the technical and business aspects of the technology industry, with a demonstrated ability to deliver.

How:
- Discover What the Customers Want;
- Drive Design, Development and Delivery;
- Act as Technical Evangelist Where Necessary.
- "Six Sigma" Methodology.

Examples:
➢ Created XXX's first OC3/OC12 FPGA, which was functional across multiple SONET-Ethernet interface 'Madonna Metro' products.
➢ Improved XXX's BMS100-Family product line integrity and robustness by 60%, achieved a cost improvement of $30M.

CAREER DRIVER

Inspiring and leading teams to develop breakthrough products that solve customer demands and have real commercial value in the global market.

SPECIAL SKILLS

My experience has honed the following development know-how:
- Execution—regularly delivering to fixed time schedules against all odds.
- Experimentation—relentless probing for new R&D and product approaches.
- Expressive clarity—strategic development plans.
- Management—optimizing people and finances to meet objectives.
- Strategic Alliances—technical liaison throughout project life cycles.
- Budget maximization—for effective use of finite resources.
- Leadership—of teams ranging from 6-30 people across multi-country-sites.

EMPLOYMENT HISTORY

XXXXXXXXX   2001–present
  Private consulting focused on bringing technology to market *"faster, better, cheaper."*

ABC Networks Limited   1983–2001
  FPGA/Hardware Designer—Maddonna Packet Switching, xxxx, 1999–2001
  Sr. Projects Leader—Technology (R&D) Organization, zzzz, 1999–94
  Projects Leader—Technology (R&D) Organization, zzzzz, 1993–90
  Task Force Leader – Quality—Customer Satisfaction, zzzzz, 1989–87
  Analyst - Quality—Reliability Engineering, zzzzz, 1986–84
  Functional Tester—Customer Service Operations, zzzzzz, 1983

EDUCATION

Project Management Professional (PMP), Project Management Institute, 2002
Bachelor of Arts, XXXX University, XXXX, Ontario, 1995
  Major: Industrial-Organizational Psychology
Electronics Engineering Technologist, XXXXXXX College, XXXXXX, XX, 1983
  Major: Telecommunications

Appendix G: Job Order Form

Nearly everything you need to know about a company is captured using this form. I say "nearly" because after the Enron debacle I believe a lot more due diligence is needed on the financials of a company to make an informed decision. I just can't help on this end. I can read a balance sheet as well as the next fellow, but if things are deliberately hidden it's impossible to tell without an audit—something most of us can not do.

I still consider my debt-to-equity question a good barometer for making a reasonable assumption about a company's ability to successfully take new products to market and therefore remain in business. So I strongly suggest you ask it.

Many of the questions can be used during the interviews as well to cross-reference people's opinion of what would make a candidate successful in a position.

◆ ◆ ◆

Job Order Form

Position Title: _____
Location: _____

Reports to: _____
Official Title: _____

Name of Assistant: _____
Assistant's direct-dial phone number: _____
Assistant's e-mail address: _____
Assistant's secure fax number: _____

GENERAL INFO:

Companies hire to solve short- and long-term problems. How quickly is a contribution expected, and what results are expected?

What is the pattern of growth or decline in the last two years? Decline is a great opportunity to turn things around financially.

What are the characteristics that the successful candidate should possess?

OVERVIEW:

Determine the reason why the job is open. Expansion? To improve on incumbent? Somebody quit or was terminated. If they want to get somebody better, then what should be added on to the last person's performance?

How many employees will this candidate be responsible for right now? In the future?

Directly? Who are they?

Who else will this candidate interact with on a daily basis? Weekly?

What should be his or her type of personality?

What should his or her leadership style be?

What should his or her management style be?

COMPATIBILITY: (fit with the job)

What are the most important functional tasks/job duties to be performed by this candidate? (What do you want this person to do?)

Are there any key responsibilities of the job that are not currently being addressed?

What are they?

What are the most important?

What is seen as the biggest challenge(s) for this position?

What are "keys to success" in the position (specific objectives in current year)?

PROFESSIONAL EXPERIENCE:

What specific work experience should the candidate have?

Will you consider other backgrounds or related experience?

What type of track record would the successful candidate have?

What companies/industries would you like to see in my candidate's background?

Are there any companies/industries that you would NOT like to see in my candidate's background?

IMPORTANT CONSIDERATIONS:

What is the background of the last two candidates hired in similar positions?

What (company/industry/country) did they come from?

Why weren't they successful here?

Why did they leave?

Where did they go?

How long where they here?

COMPENSATION:

➢ Find out about compensation and benefit RANGE. Don't be concerned with specifics. Are you compatible with the range? There is no cutoff point. Your skill in interviewing will maximize what you get and stretch the range.

You must show that your value is more than your cost.

What is the expected first-year compensation?

Base salary range?

Desired hiring base salary

Expected annual bonus?

Stock options

What is the growth potential/expectation (in terms of position and/or earnings)?

Salary growth anticipated

Ideal starting salary?

Absolute highest $ for a top candidate?

Position growth anticipated in the next 12 to 24 months?

Fringe benefits?

Medical?

Retirement?

Time-off policy?

◆ ◆ ◆

Appendix H: Organizer

This organizer is designed to plan and monitor all the essential components of a job search. It's meant to be taken to your local copy shop and copied on to an 11" x 17" sheet of paper. This becomes your blotter for your desk.

You start on the left-hand side every morning with networking calls to friends/associates who you've sent your resume to. The easiest way you are going to find a job is through your network, so start there. Put the effort where the results are going to appear first. You also want to do this first thing in the day because generally you won't have been rejected by many people yet, so your voice will be filled with enthusiasm as you speak to people on the phone.

Call the recruiters next. Most recruiters block out from 7 to 10 a.m. for marketing calls so, although you want to get them early in the morning before they have time to forget who you are, you don't want to interrupt their marketing calls. They could be marketing you, after all, and you don't want to remind them they haven't found anything for you yet and throw off their concentration.

Next, go into your calls to those companies you identified yesterday as potentials.

Lastly, make a list of companies for tomorrow and start researching them on the Web.

◆ ◆ ◆

Networking Calls

Date: _____

Follow-up Time	Person	Phone Number	E-mail	Referred

Agencies/Recruiters

Company	Person	Phone Number	E-mail	Action

Web and Newspaper Ads

Company	Hiring Manager	Position	Phone Number	E-mail	Action

Weekly Goals

Networking calls: _____

Agency calls: _____

Ads: _____

Targeted Research: _____

Interviews This Week:

_____

_____

_____

_____

Prospects & Target Research (cold calls)

Company	Website	Contact	Phone #	E-mail	Comments

Weekly Activity Plan

(Objectives/Performance)	WEEK 1	WEEK 2	WEEK 3	WEEK 4
Prospect/Cold Calls				
Networking				
Contact Search Firms/Agencies				
Answer Ads				
Information Interviews				
Thank You Notes				
Research and Read				
Job Interviews				
TOTALS				

Appendix I: Sample Behavior Interview Questions

The following IT-related behavioral questions are pulled straight from the database of behavioral questions at the Software Human Resource Council (reprinted here with permission.). You would be well advised, if you're looking at IT positions, to check them out the Council at *www.shrc.ca.* Its database is the first of its kind in the world and widely used by IT organizations in the United States, Canada, India, China, Egypt, and Europe.

•••

Technical/Functional Skills

Design IT Strategy or High Level Architecture

Describe a time when you successfully developed the IT strategy of an organization.

* Describe the planning that went into setting the strategy.
* What factors did you take into account when setting this strategy?
* How did you communicate this strategy to others in the organization?
* How successful was this strategy in driving the business to success?

Can you think of a time when you had to define an enterprise level data model or architecture?

* Describe the situation.
* What steps did you follow in defining this model/architecture?
* What was the outcome of the situation?

Manage the Knowledge Infrastructure

Have you ever had to develop data, process, or network models for your organization?

* What elements did you consider in coming up with an appropriate model?
* How did you analyze these elements in order to determine the "best fit"?
* What was the outcome?

Develop New Software and Databases

Describe a time when you were required to perform high-level requirement analysis for other software developments and IT professionals.

* Who were you defining the requirements for?
* What input did you seek and from whom?
* Describe your methodology in performing the analysis.
* Do you think you were successful? Why?

Maintain Existing Software Applications and Databases

Describe how you have used different software to enhance your (or a client's) organization's overall effectiveness.

◆ Provide a particular example of when you have done this.

◆ What software did you use?

◆ How did you know that this particular software would be an asset to the organization?

◆ What was the outcome?

Operate the Delivery Platform (processors, networks, LANs, WANs)

Describe your approach to monitoring the operations of a delivery platform.

◆ Please provide a specific example.

◆ What metrics did you use?

◆ What types of problems did you encounter?

◆ What resolutions did you propose?

◆ How do you know that you were successful in resolving the problems?

Financial Management

Give an example of a time in which you felt that you were successful in preparing a sound financial plan.

◆ Describe the circumstances.

◆ Describe your methodology.

◆ Which unknowns were you dealing with?

◆ How did you plan for these unknowns?

◆ Why do you feel that you were so successful in your task?

How do you ensure that the costs associated to your projects and activities are in line with budgeted amounts?

◆ Can you think of a specific project or activity as an example?

◆ What monitoring mechanisms did you utilize?

◆ What unforeseen events did you encounter that forced you to reassess your planned expenditures?

◆ How did you compensate for those unforeseen events?

◆ What was the outcome?

People Management

Have you ever had the responsibility of defining organizational structures and jobs?

◆ What was the situation?

◆ What information did you take into account when defining the appropriate structure or job?

◆ What stakeholders did you involve in your decisions?

◆ What impact did your decisions have on the organization and its viability?

When recruiting, selecting and firing employees, how do you determine who matches job requirements and who does not?

* Can you give me an example of a time when you did this?
* What positions were you recruiting for?
* How did you define the job requirements?
* What assessment tools did you utilize?
* How successful was the new incumbent in the position? Explain.

IT Management

Describe how you have used an understanding of IT business management to further the organization's goals.

* What is a specific example?
* What challenges did you face?
* How did you overcome these challenges?
* What was the outcome?

Leadership

What have you done in the past to inspire others to own the organizational vision?

* Provide a specific example.
* Describe the vision.
* What methods of communication did you use to articulate the vision?
* How did you inspire others to own that vision?
* What was the result of your efforts?

Provide an example of a time when you set an example for others by behaving in a manner that supported the organizational goals and values.

* Describe the circumstances.
* What behaviors did you want others to model?
* How did these behaviors support the goals and values of the organization?
* How do you know that you were successful in setting an example for others?

Project Management

Describe a recent project that you were involved during which you estimated and planned timelines and milestones.

* Describe the project.
* What estimating and planning strategies did you use?
* How accurate were your estimated timelines/milestones?
* Looking back, what would you have done differently?

Vendor Relations Management

Describe a challenging situation in which you had to write technical specifications.

• For what purpose were you writing these specifications?
• What challenges did you encounter?
• How did you overcome these challenges?
• What was the outcome?

Describe a particularly challenging RFP that you were required to prepare.

• Describe the circumstances.
• Describe your approach to writing the RFP.
• What things did you take into consideration when preparing the RFP?
• What were the results of your actions/decisions?

Client Relations Management

Describe a situation in which you successfully negotiated the terms of a win-win client contract.

• What were you negotiating and for what purpose?
• Describe your negotiation tactics.
• Did the contract that you negotiated fulfill its objective? How?
• Why do you believe that the terms of this contract were win-win?

Personal/Interpersonal Skills

Working as an Individual

Scanning the environment in order to extract relevant information is a critical component of this job. Discuss a recent environmental scan that you performed, and describe the relevance of the information gathered through this scan.

• What was the purpose of the environmental scan?
• What data collection methodologies did you utilize?
• What types of information did you obtain?
• How did you determine the relevance of the information?
• What was the result of the scan? / What was the information used for?

In any work environment, there exists both written and unwritten codes of conduct and service standards for conducting business. Describe how you have adhered to these codes of conduct/standards to maintain professionalism in your business dealings.

• Provide a specific example of a situation where it was difficult to adhere to these codes of conduct or standards.
• What codes of conduct/standards were relevant to this situation?
• Why was it difficult to adhere to them?
• What steps did you take to adhere to them?
• What was the result of your actions/decisions?

Communications

Provide an example of a time when you listened, then responded to the ideas or concerns of another individual.

+ Describe the circumstances.
+ What information was this individual trying to communicate?
+ What did you do to ensure that you understood this oral message?
+ How did you respond to this individual?
+ Describe the outcome of the situation.

These non-IT related questions are likely to be asked as well, with the same rigor of drill-down from the interviewer.

1. In viewing your candidacy for this position, in what areas do you feel you would be a particularly strong performer? Why?
2. Describe your three greatest strengths and tell me how you used them to bring about improvements in your job.
3. In what ways could you enhance your overall professional capability?
4. What aspect of your overall qualifications has most stood in the way of your career advancement to date?
5. Based on how you feel your boss would rate your current job performance, what areas would he or she cite as:
   + Exceeding expectations? Why?
   + Meeting expectations? Why?
   + Falling short of expectations? Why?
6. If you sense you are not fitting in well with a group and feel that you are being treated as an outsider, what would you do?
7. What examples can you cite that best demonstrate your ability to relate well to others?
8. What type of work do you find most satisfying from a professional standpoint?
9. What type of work do you find least satisfying from a professional standpoint?
10. What have you learned from your past work experience about the type of work you ost enjoy doing?
11. Which of your past work environments came closest to your "ideal"?
12. Which of your traits and characteristics do you find most frustrating?
13. Of which of your personal traits and characteristics are you most proud, and why?
14. Describe the planning and decision-making processes important to successful business operations.
15. In your opinion, what do unsuccessful organizations planning and decision-making processes look like? What are they missing?
16. Describe your overall planning process.
17. What are some of the techniques you use to motivate poor performers?
18. Give me an example of something very creative that you did.
19. What is, perhaps, the most complex business analysis you have had to make?

20. What factors made it complex? Give me an example of a complex communications problem that you faced.
21. What made it complex? If your boss told you that you had a "stupid idea," but you knew it was a very good one, what would you do?
22. Someone in another department, with whom you have infrequent contact, has been saying some uncomplimentary things about you; what would you do?

Appendix J: Accounting Job Boards

Because accountants and finance people are the backbone of every high-tech company we've included accounting-only job boards for you. As a serious job seeker it's a good idea to surf these boards to see who is hiring and growing in your area.

◆◆◆

Actuarial Actuary: *www.actuary.com*
Actuarial Actuarial Jobs: *www.sitepowerup.com/mb/view.asp?BoardID=106435*
Actuarial Financial Job Network: *www.financialjobnet.com*
American Association of Finance and Accounting: *www.aafa.com*
Auditor Financial Job Network: *www.financialjobnet.com*
Commerce Global Career in Commerce: *www.globalcareers.com/commerce/*
Comprehensive Accounting and Finance Jobs: *www.accountingjobs.com/*
Comprehensive Accounting Prof: *www.accountingprofessional.com/*
Comprehensive Accounting.com: *www.accounting.com/*
Comprehensive American Institute of CPAs: *www.aicpa.org/*
Comprehensive eFinancial Jobs: *www.efinancialjobs.com/*
Comprehensive Jobs in Accounting: *ww2.itoday.com/jobs-careers/banking/*
Comprehensive MBA Careers: *www.mbacareers.com/*
Comprehensive MBA Employment Connection: *www.MBAnetwork.com/meca/*
Comprehensive MBA Jobs: *www.mbajob.com/*
Comprehensive National Black MBA Association: *www.nyblackmba.org*
Comprehensive National Business Employment Weekly: *www.nbew.com/*
Comprehensive National Society of Accountants: *www.nsacct.org/*
Comprehensive Pro2Net Accounting: *www.pro2net.com/*
CPA American Institute of CPAs: *www.aicpa.org/*
CPA MBA Careers: *www.mbacareers.com/*
CPA MBA Employment Connection: *www.MBAnetwork.com/meca/*
CPA MBA Jobs: *www.mbajob.com/*
CPA National Black MBA Association: *www.nyblackmba.org*
CPA American Association of Hispanic CPAs: *www.aahcpa.org/*
CPA Financial Job Network: *www.financialjobnet.com*
Nationjob: *www.nationjob.com*
www.accountantjobs.com/
www.accountingjobs.com/index.html
www.rutgers.edu/Accounting/raw/aaa/placemnt.htm
www.jobcontrolcenter.com/ccn/nabainc/
www.accounting.com/employment/

Appendix K: Professional Marketing Associations and Institutes

Because nothing moves until marketing packages the opportunity, your skills are in high demand everywhere.

◆◆◆

Academy of International Business: *aib.msu.edu/*
Academy of Management: *www.aomonline.org/*
Academy of Marketing Science: *www.ams-web.org/*
Advertising Research Foundation: *www.arfsite.org/*
American Academy of Advertising: *advertising.utexas.edu/AAA/*
American Association of Advertising Agencies: *www.aaaa.org/*
American Collegiate Retailing Association:
    www.sba.muohio.edu/acra/ACRA%20files/ACRA%20Constitution.htm
American Management Association: *www.amanet.org/index.htm*
American Marketing Association: *www.marketingpower.com/*
American Psychological Association: *www.apa.org/*
American Statistical Association: *www.amstat.org/*
Asia Pacific Marketing Federation: *www.apmf.org.sg/*
Asociacion Española de Marketing Directo: *www.aam-ar.com/socios/elegidos.htm*
Association for Consumer Research: *www.acrweb.org/*
Association Française du Marketing *www.adetem.org/*
Association of Management / International Association of Management:
    www.aom-iaom.org/index.html
Association of Marketing Educators: *www.cwu.edu/~luptonr/mea/*
Atlantic Marketing Association: *www.atlanticmarketing.org/*
Australia-New Zealand Marketing Academy: *www.anzmac.org/*
Business Marketing Association: *www.marketing.org/common/home.asp?ShowType=bma*
Canadian Marketing Association: *www.the-cma.org/*
Center for Service Marketing: *www.itsma.com*
Chartered Institute of Marketing: *www.cim.co.uk/cim/index.cfm*
Classification Society of North America: *www.pitt.edu/~csna/csna.html*
Direct Marketing Educational Foundation: *www.the-dma.org/dmef/index.shtml*
DocSIGi: *docsig.eci.gsu.edu/*
E-Marketing Association: *www.the-ema.com/*
ESOMAR: *www.esomar.nl/*
European Academy of Management: *euram.iese.edu/*
European Institute for Advanced Studies in Management: *www.eiasm.be/index1.html*
European Marketing Academy: *www.emac-online.org/associations/emac/index.asp*
FEDMA (Federation of European Direct Marketing):
    www.ftc.gov/bcp/icpw/comments/fedma.htm
Hong Kong Institute of Marketing: *www.hkim.org.hk/*
Hospitality Sales and Marketing Association International: *www.hsmai.org/*

Institute for the Study of Business Markets: *www.smeal.psu.edu/isbm/*
Institute of Canadian Advertising: *www.ica-ad.com/*
Institute of Public Relations: *www.ipr.org.uk/*
International Advertising Association: *www.iaaglobal.org/default.asp?subsiteID=2*
International Association for Management of Technology: *www.iamot.org/*
International Federation of Classification Societies: *www.classification-society.org/*
International Newspaper Marketing Association: *www.inma.org/*
Internet Marketing Association: *www.iimaonline.org*
Japan Marketing Association: *www.jma-jp.org/eng/jheng.htm*
Market Research Society: *www.mrs.org.uk/*
Marketing Associatie Nederland: *www.ma-nederland.nl/*
Marketing Institute of Ireland: *www.mii.ie/*
Marketing Institute of Singapore: *www.mis.org.sg/*
Marketing Management Association: *www.mmaglobal.org/home.html*
Marketing Research Association: *www.mra-net.org/*
Marketing Science Centre at University of South Australia: *www.marketingsciencecentre.com/*
Marketing Science Institute: *www.msi.org/*
Pi Sigma Epsilon: *www.pisigmaepsilon.org/*
Product Development and Management Association: *www.pdma.org/*
RITIM (Research Institute for Telecommunications and Information Marketing): *ritim.cba.uri.edu/*
Sales and Marketing Executives International: *www.smei.org/*
Sales Special Interest Group: *mkt.cba.cmich.edu/salessig/*
SMT (Professional Society for Sales and Marketing Training): *www.smt.org/*
Society for Marketing Advances: *www.marketingadvances.org/*
Society for Marketing Professional Services *www.smps.org/*
Strategic Management Society: *www.smsweb.org/*
The Direct Marketing Association: *www.the-dma.org/*
The Society for Consumer Psychology: *fisher.osu.edu/marketing/scp/*

— Appendix L: Internet Search Engine Cheat Sheet —

Here are the major rules for using the largest search engines to source hidden jobs. Just select the key words for your chosen job, plug them in, and go! Remember that Google is the largest but that it doesn't cover the whole Internet, so use a few others as well.

◆ ◆ ◆

Search Engine	Job Searches	Boolean Commands	Membership Directories	Alumni Searches
Google www.google.com Advanced Search	Keyword Keyword Keyword (inurl: job OR title:job)	AND *(default)*, OR. + *only* to include stop words *or* to require specific domain in URL or on page (e.g., +edu). - to remove.	Keyword keyword (directory OR contact (inurl:member OR intitle:member)	Companyname keyword Archive (inurl:list OR intitile:list)
AltaVista www.altavista.com Search Assistant *(Assisted Basic Search)* Advanced Search *(best to use)* Subject Directory *(LookSmart)*	(keyword AND keyword AND keyword) AND (url:job OR title:job)	+ to require. - to remove. OR *(default)*. **Advanced Search only: AND, OR, AND NOT, NEAR,** *Nesting* ( ); **Sort by: terms here activate results ranking.**	keyword AND keyword AND (directory OD contact(AND (url:member OR title:member)	companyname AND keyword AND (directory OR contact) AND (url:member OR title:member)
Northern Light northernlight.com Power Search No Subject Directory	keyword AND keyword AND keyword AND job	+ to require. - to remove. AND *(default)*, OR, NOT. Nesting ( ). **Stems** plurals.		
HotBot hotbot.lycos.com Advanced Search Subject Directory *(Open Directory)*	keyword AND keyword AND keyword AND job	+ to require. - to remove. AND *(default)* (all of these words); OR (any of these words); NOT: and PHRASE *(can also use quotes)* **Select Boolean** and can *nest* ( ).		

Appendix M: *How You Hire*

2003 North American Survey of High-Tech Employers
Final Report September 2003

Introduction

The North American technology industry crashed during the first quarter of 2001. A recent study by the American Electronics Association shows High-tech manufacturing employment declined by 415,000[1] over the past two years. Another study by the Information Technology Association of America shows 500,000 IT[2] jobs were also lost. Canadian job loss numbers were on scale with the US. In the first 18 months more than 1 million jobs were eliminated in the U.S., and an estimated 100,000 jobs in Canada. These jobs are the ones at the forefront of the Knowledge Economy.

While Telecom was the hardest hit, no industry escaped unaffected. The Tech-Wreck that followed the burst of the DOTCOM bubble was devastating. Just as many highly qualified high technology professionals are experiencing difficulty finding employment, employers across North America continue to report shortages for highly skilled workers[3]. What's going on?

Two key questions emerge that must be answered to allow us to understand the conflicting signals. The first, for those workers who are now actively looking for work, is: Do the traditional ways to look for a job still work? In other words, do employers in different tech sectors look for different things?

The second question, for executives trying to find highly-skilled workers who are in short supply, is: Do Executives look for the same things in a new hire that Non-Executives do? Do Human Resource staff interview for the same skills Engineering does?

Are there new and better ways to find a job, and to find a candidate?

Sensing a shift in the way employers were hiring Perry-Martel International Inc., in conjunction with the Canadian Advanced Technology Alliance (CATA), set out to survey North American employers on their actual hiring practices and develop a guide to tell technology workers how to best find new employment in the post DOTCOM high-tech economy.

The survey looks at how hiring managers differed between:

◆ The United States and Canada;
◆ Major geographic locales within the USA and Canada;
◆ Major segments of the high-technology industry; and
◆ Titles or positions.

The Survey results turn the old "networking" networking" networking" adage on its head.

---

1 Tech Employment Update, AEA, *www.aeanet.org/publications/idmk_endofyear2002.asp*
2 2003 Workforce Survey, Information Technology Association of America, *www.itaa.org/workforce/studies/03execsumm.pdf.*
3 Attracting and Retaining IT Talent in a Dynamic Economy: The 2002 High Tech Worker Survey Report, IBM Consulting Services Canada, *www.itworldcanada.com/multimedia/pdfs/high-tech-workers-survey.pdf.*

Methodology

The survey findings are based on feedback provided from "hiring managers" who are currently employed in the high-technology industry across North America. Feedback was provided through a self-completed 5-question survey.

Nationality

The distribution of participants across North America was approximately 20% lower in the United States than in Canada. We believe this is due to the fact the database of contacts contained more Canadian content. *(Figure 4)*

Geography

The United States was divided into East Coast USA, West Coast USA, and Central USA. Participants were grouped into major Canadian cities; Montreal, Quebec, Ottawa, Toronto, Ontario (other) Calgary/Alberta, and as well as Eastern Canada and West Coast Canada. (Figure 5 and Figure 6)

Occupational Areas

Participants were asked to self select their titles which were divided into executive and non-executive. The executive roles were: CEO/President, VP Sales, VP Marketing, VP Engineering, VP Human Resources, and Executive. A majority of respondents who signified "Executive" where in the finance area. Non-executive roles consisted of Sales, Marketing, Engineering, Finance, and Human Resources. *(Figure 8 and figure 9)*

Industries of Employment

Industrial classifications included Software, Hardware, Telecom, Biotech, Government, and Other. The majority of respondents in the "Other" category are Service companies (73%) and Medical Device companies 17%. The Biotech category also included companies, which could be regarded as Biometrics. *(Figure 7)*

Sample Frame

The sample frame for the survey was a random sample of contacts from the databases of Perry-Martel and CATA. Contacts were from software, hardware, biotech, telecom, government, and other technology industries.

Sample Size

A total of 19,000 contacts were invited to respond to the survey. A total of 7182 people completed the survey for an overall response rate of just over 37.8%. We attribute the unusually high response rate to the brand awareness of the CATA. Its news bulletins are widely read throughout the technology industry in the United States and Canada. The final survey sample was considered to be largely representative of the database.

Survey Delivery

The survey was conducted exclusively on the Internet by emailing each contact a direct link to the survey which was hosted on CATA's web site. Each contact's email address was used as means to provide a unique PIN (Personal Identification Number) to control the sample group and ensure the integrity of the data.

Survey Questions

Participants were asked the following two questions:

- Question 1: What is the best way to find a high-tech job today?; and
- Question 2: "IF" you were going to hire a new employee in today's economy, what qualities or characteristics would you be looking for?

Question 1 is based, on the, "10 Ways to Find a Job" as anecdotally defined by Richard Boles in his best selling book, **What Color is Your Parachute**[4].

The question and possible answers were:

What is the best way to find a high-tech job today?

1. Randomly mail out resumes to employers.
2. Target Marketing to companies that you've researched who have a problem you can solve.
3. Answer ads in professional or trade journals appropriate to your field.
4. Answer ads in newspapers in other parts of the province/state/country.
5. Answer ads in local papers.
6. Contact headhunters or personnel recruitment firms.
7. Networking.
8. Knock on doors in a business park or office building.
9. Call companies from the Yellow Pages.
10. Job Hunter's Club

First written nearly 30 years ago and updated yearly, **What Color is Your Parachute** is the definitive resource for job seekers.

Question 2 was an open-ended free form question where participants were able to say whatever they wished.

Survey Findings

Q1: What is the best way to find a high-tech job today?

Executives prefer a direct approach. Executives expect Job-Seekers to have researched their company and be able to tell them specifically how they can add value. Among Executives 97% thought Target Marketing to companies was the best way to find a new job, regardless of the seniority of the job being sought.

It is interesting to note that only 31 Executives mentioned Job Boards as an effective recruiting tool, or pointed out the web was not a choice in Question 1. It would appear that Job Boards are not as effective as many may have thought—a conclusion backed by a Forrester Survey which found that only 4% of people had secured jobs using Job Boards.[5]

---

4 _What Color Is Your Parachute 2003: A Practical Manual for Job-Hunters_, Richard Boles, Ten-Speed Press.
5 Forrester

Executives

Regarding "the best way to find a high-tech job today":

➤ Statistically there were no differences between the United States and Canada on "the best way to find a high-tech job today".

➤ There were no differences between industries (software, hardware etc.) on "the best way to find a high-tech job today."

➤ There were no differences among different types of executives (sales, marketing, CEO, HR, Engineering) about the best way to find a high-tech job today.

➤ Among Executives **97% feel the best way to find a high-tech job today is: Target Marketing** to companies that you've researched, which have a problem you can solve.

➤ The second best way to find a job is: Contact headhunters or personnel recruitment firms at 1.9%

➤ Networking rated a 0.65%

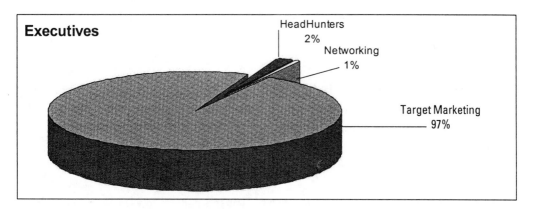

Figure 1 Best Way to Find a Job - Executive

Non-executives

Among Non-executives, 73% felt that Networking was the best way to find a job. This was highest among those who were in a Human Resource position – 83%, while Engineering types favoured Head-hunters.

Non-executives are obviously different from Executives:

Regarding "the best way to find a high-tech job today":

➤ Statistically there were no differences between the United States and Canada on "the best way to find a high-tech job today".

➤ There were no differences between industries (software, hardware etc.) on "the best way to find a high-tech job today."

➤ Among Non-Executives 73% feel the best way to find a high-tech job today is Networking.

➤ Human resource professionals favored Networking - 83%.

➤ Engineering Professionals favored Headhunters – 57%

➤ The second best way to find a job is: Contact headhunters or personnel recruitment firms at 22.9%

➤ The third best way to find a job is: Target Marketing at 1.8%

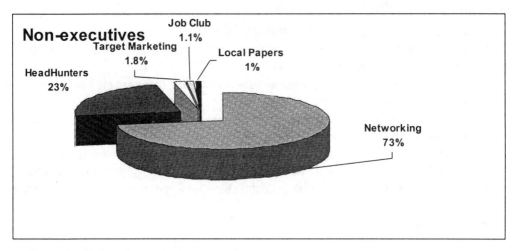

Figure 2 Best Way to find a Job - Non-executive

Survey Findings

Q2: IF you were going to hire a new employee in today's economy, what qualities or characteristics would you be looking for?

Once again the differences are most apparent among Executives versus Non-executives, not—as one might think—between industries, nations or regional geography.

Executives

It is interesting to note that 31% of Executive participants had a fellow Non-executive participant from the companies who completed the survey. We had not anticipated this cross-pollination of participants. The most interesting finding was how polarized their hiring indices were. Executives indicated (84%) that they were in interviewing/hiring mode versus 1% of non-executives in the same firm versus 7% of Non-executives over all. Perhaps execs and non-execs are out of sync because execs are always in strategic hiring mode and non-execs are not.

There appeared to be no real differences between the United States and Canada on what they were looking for in a new hire. Nor is there a difference between industries or geography, although many US firms on the West Coast – 33%, specifically mentioned "loyalty".

What's most interesting is that there was **no indication** that executives from different departments interview for different things. Engineering versus Sales). Nor were there any indications they were looking for different qualities/characteristics between executives versus non-executives. I think it would be fair to conclude that the execs are looking for certain CORE CHARACTER-ISTICS or QUALITIES in each candidate? Execs are focused on those qualities which" add value", "revenue" and "wealth" to the company in short order. Executives concentrate almost entirely on intangible skills instead of tangible skills.

> ➤ Across the spectrum of executive positions, 81% focused on intangibles. Here are the most often mentioned characteristics/qualities they were looking for: leadership, initiative, drive, entrepreneurial spirit, dedication, creativity, and loyalty
> ➤ **Wealth creation was the single most desirable outcome mentioned** (83%).

➢ There are no differences between industries with the exception of "Other" where service companies were looking for good "client relationship" skills or an "outgoing personality".

➢ **53% indicated anecdotally they would interview/hire exceptional candidates.**

Here is our summary of some the top characteristics/qualities we noted Employers now value more than tangible hard skills. We are calling this the New Value Table:

The New Value Table

An Employer's Value Requirements	A Job-seeker's Quality That Counts
Create new intellectual wealth for my company; add to my intellectual assets.	A consuming desire to make something new; to cut a new path rather than take a road.
High-energy enthusiasm for the job, regardless of the hours worked.	Work is a game—an integral, vibrant part of his or her life.
Not only is money not the most important issue—it's beside the point.	Internal pride to leave a "legacy signature" on their work, rather than strive for a paycheck.

Typical Executive's Response to Question 2:

"A degree of self confidence together with a keen desire to succeed, no matter what the obstacles. We're always looking for exceptional players."

Non-executives

There appeared to be no real differences between the United States and Canada on what they were looking for in a new hire. Unlike Executives who focused on intangible qualities/characteristics, Non-executives concentrated almost entirely on tangible skills and/or experience.

Non-executives will only hire someone when the Job-seeker's skill set is an exact match for their spec. Engineering on the other hand appears more willing to hire people who can learn new technologies quickly and have the ability to think through problems.

➢ The software industry on both sides of the border stressed "education" - 47% and "references" - 55% of the time.

➢ Among American companies 67% stressed references, versus 7% in Canada.

➢ Across all positions 77% of Non-Executives focused on tangibles, identifying specifics such as C++, JAVA, E-commerce, advanced education, previous experience, and references.

➢ 87% were looking for "team players" or candidates with "people skills".

➢ 7% indicated they would interview/hire exceptional candidates.

Non-Executives are still looking to fill in boxes on a standard employee recruitment form.

Typical Non-Executive's Response to Question 2:

"Has the ability to write source code that is semantically clear to the human reader. Takes pride in quality of work. Intelligent. A team player. Detail-minded."

Discussion

Participants differed on the best way to find a job according to whether they were in an Executive or Non-executive position.

Executive participants ranked Target Marketing as the best way to find a high-tech job today (97%) and Head-hunters at 1.9% and Networking at 0.65%. In stark contrast, Non-executives ranked Target Marketing at 1.8% Head-hunters at 22.9% and Target-Marketing at 1.8%.

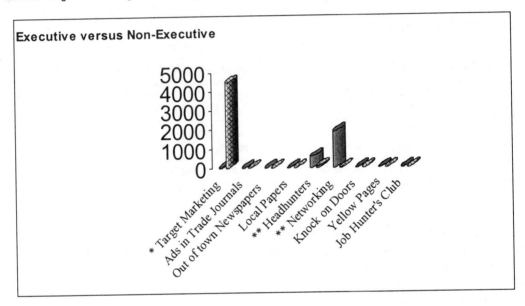

Figure 3 Executive versus Non-executive Comparison

The biggest surprise in the survey was that Networking rated so low among Executives at 0.65% but rated 73% with Non-executives. Executives prefer a direct approach and are less likely to suggest using Head-hunters, where as Non-Executives still believe Networking is the best way to find a job.

These differences provide the answers to the two key questions posed earlier, for have serious Job-seekers and Employers.

The first question, for Job-seekers, is: Do the traditional ways to look for a job still work? The immediate activity that happens in a recession is layoffs. This usually coincides with a hiring freeze. When this occurs no one outside the executive suite (CEO, President, VP) has the authority to hire new employees – full stop. This being the case, it's easy to understand why Non-executive staff may be loath to bring forward anyone they interview no matter how strong their skill set.

In other words, the traditional approaches to finding a job do not work! The Networking favoured by Non-executives will keep Job-seekers very busy but is not likely to result in a job offer. Activity is not the same as results. Moreover, 90% of Head-hunters work with Non-executive hiring managers and rarely speak with Executives, which means that a Job-seeker's desire to find a placement through a Head-hunter is also unlikely to succeed unless they are working on a specific project for which they are an exact fit for. This effectively eliminates 95% of the options favoured by Non-executives and promoted by the current crop of career management books.

In answer to the second question — what is the best way for Executives to find highly-skilled people — the best way – perhaps the only way – to obtain the ideal employee is the direct way. Employers assess a candidate's "value" in directly targeted searches, where the value proposition of intangible characteristics is tracked by the Executive — the one person who is empowered listen to the candidate's case directly.

There is a Catch 22 in all of this, while Executives are looking for "qualities" and "solutions" (people who can advance their business, make them more money, save them money, or increase their efficiency) they have tied the very hands of the people who are most likely to meet their next start candidate with a hiring freeze.

It is only wishful thinking that this will change anytime soon because the tech industry is still very much in a reactive mode.

Observations

The North American tech-community has seen more change in the last twenty-four months than in the past ten years. Billion dollar companies – gone! And not just one or two... We are talking about scores of them – bankruptcies, mergers, acquisitions and total retrenchments. The right people are more important to companies than ever before and the rules for hiring have changed.

For job seekers, heavy preparation and due diligence are necessary to find which companies have an opportunity for them recognizing that, especially for its senior positions, a high-tech company is rarely looking to check-off a box on a standard employee recruitment form. They are looking for a senior person who can deliver a Quality not a quantity. It's up to the job-seekers themselves to clearly articulate their unique value based on the VALUE an employer is seeking. Obviously approaching employers in a targeted manner reaps the best results – it's also what Executives expect.

Those companies struggling to get through the impacts of the high tech meltdown are guiding their search for the top people whose skills make the difference between victory or defeat very cautiously. Its shareholders are demanding Qualities are difficult to find, measure or test in an ordinary recruitment drive. And you don't find those qualities by searching for specific salary levels — the qualities that make up the new Value Table are money-resistant.

Resolving the Problem

In this leadership – hungry environment, recruiting and looking for a job have changed. There is still a high demand for people who can design a top product, manage complex projects, perform marketing miracles, sell new customers, or execute on a plan. To resolve the supply demand imbalance both sides need to approach the problem differently.

Job-seekers need to become adept at self-promotion — learn how to search the world, cold-call prospects, get their attention, raise your proposition above the background noise, keep at it tenaciously for however long it takes – be it weeks or months – and be intelligent enough to present your skill set in creative new lights until the persuasion works. Before approaching an employer, the first thing a job-seeker needs to ask himself is, "How can I increase shareholder value? Answering that question and reinforcing their value message to prospective employers will secure them a job faster than the tired - old fashioned - networking promoted by career books today.

Employers on the other hand, need to recognize that value is not salary; worth does not flow from a job title. Knowing how to evaluate the worth of someone's contribution is the important element. Being able to assess a candidate's star performance capabilities defines the difference between average and extraordinary. Employers need to be in strategic hiring mode at all times – constantly assessing how a new hire might add value for their shareholders. Most importantly they need to ensure their middle managers are in sync with the company's vision and the new people, which are required to execute on the strategy.

Appendix

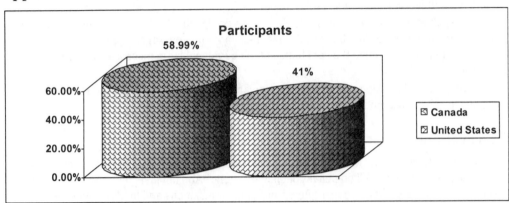

Figure 4 Participants by Nationality

United States

The technology sector in the United States touches virtually every state. For simplicity sake we divided the United States into Eastern USA, Central USA, and Western USA. Participants were asked to self-select the region they were living in. We did sampling with in California and Washington State to see if there were any major differences in the voting patterns and there were not. We are comfortable stating the results are representative of each region.

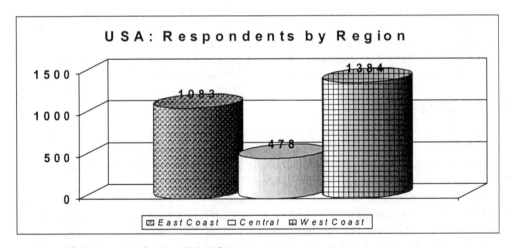

Figure 5 Participants by Region - United States

Canada

Canada, because of its population size and specific geographic concentrations of technology clusters, was more finely aggregated into: Eastern Canada, Montreal, Quebec city, Ottawa, Toronto, Ontario Other, Calgary/Alberta, and West Coast. Once again a sampling was done in the regions off Ottawa, Toronto, and Ontario Other to see if there were any major differences in the voting patterns and there were not. We are comfortable stating the results are representative of each region.

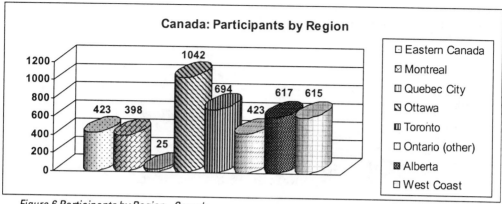

Figure 6 Participants by Region - Canada

Industry

We aimed to get as broad a cross section of the technology industry as was reasonably possible but kept the aggregation to Software, Hardware, Telecommunications, Biotech Government and Other.

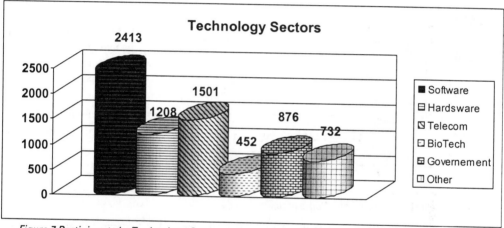

Figure 7 Participants by Technology Sector

Participants by Title

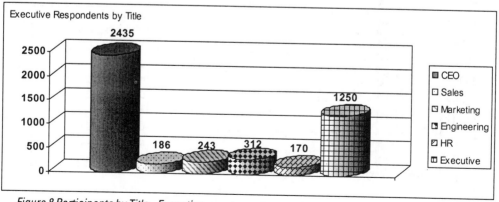

Figure 8 Participants by Title - Executive

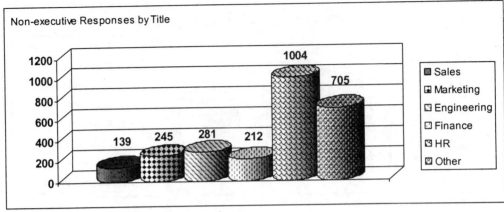

Figure 9 Participants by Title - Non-executive

Participants by Preference

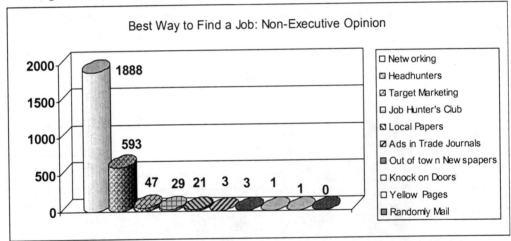

Figure 1 Best Way to find a Job - Non-executive

Executive Participants ranked Target Marketing as the most effective way to get a job at 97% and Head-hunters at 1.9%.

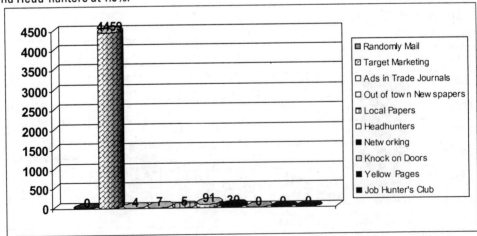

Figure 2 Best Way to find a Job – Executive

Appendix N: Associations

The following technology associations are a bounty of resources and leads. Many of them have their own job boards.

It's been my experience that association executives are all very approachable. It's in their best interest to help their member companies grow.

Always drop the people you talk to here a thank you note and a copy of your resume electronically. People remember referrals generally 15 minutes after they get off the phone, so sending a thank you note and electronic resume can pay huge dividends.

As well, I have included links to technology event organizers that employers frequent and links to hard-core research sites to get the latest on technology and events. Most of the job growth over the next few years will be in small companies of five to 20 people who are under the radar of the press and most job seekers.

◆◆◆

Associations

American Engineering Association *www.aea.org*
Chicago Software Association (CSA) *www.csa.org/*
Colorado Software Association (CSA) *www.csadirectory.org/default.asp*
Computer Measurement Group (CMG) *www.cmg.org/*
Computing Research Association (CRA) cra.org/
Data Management Association (DAMA) *www.dama-ncr.org/*
European Information Technology Services Association (EISA)
    www.worldcongress2000.org/en/affiliates/eisa/eisa.html
Federation On Computing in The United States (FOCUS) *www.acm.org/focus/*
Greater Cincinnati Software Association (GCSA) *www.gcsa.org/*
Independent Computer Consultants Association (ICCA) *www.icca.org/*
Informatics Society Of Iran (ISI) *www.gpg.com/MERC/org/isi/isi.html*
Information Systems Consultants Association (ISCA) *www.isca.org*
Information Technology Association Of America (ITAA) *www.itaa.org/*
Institute For Certification Of Computing Professionals (ICCP) *www.iccp.org/*
International Association Of Computer Professionals (IACP) *www.pros-n-cons.com/*
International Council For Computer Communication (ICCC) *www.icccgovernors.org/*
Software & Information Industry Association (SIIA) *www.siia.net/*
International Software Process Association (ISPA) *www.ece.utexas.edu/~perry/prof/ispa/*
Internet Society (ISOC) *www.isoc.org/*
International Association For Mathematics and Computers In Simulation (IMACS)
    www.cs.rutgers.edu/~imacs/
Information Systems Audit and Control Association (ISACA) *www.isaca.org/*
Maine Software Developers Association (MESDA) *www.mesda.com/*
Object Management Group (OMG) *www.objs.com/survey/omg.htm*
San Diego Software Industry Council (SDSIC) *www.sdsic.org/*
Society For Information Management (SIM) *www.simnet.org/*
Society For Software Quality (SSQ) *http://ssq.org/welcome_main.html*

Software Forum (SF) *www.softwareforum.org/*
Software Productivity Centre (SPC) *www.spc.ca/*
Special Interest Group On Measurement & Evaluation (SIGMETRICS) *www.sigmetrics.org/*
Video Software Dealers Association (VSDA) *www.vsda.org/*
Washington Software Alliance (WSA) *www.wsa.org/*
World Information Technology And Services Alliance (WITSA) *www.witsa.org/*
American SAP Users Group (ASUG) *www.asug.com/*
Association of Information Systems Professionals (DPMA) *http://aisp.bus.wisc.edu/*
Institute of Electrical and Electronic Engineers (IEEE) *www.ieee.org/*
The USENIX Association (USENIX) *www.usenix.org/*
Worldwide Institute of Software Architects (WWISA) *www.wwisa.org/*
ACM Special Interest Group in Software Engineering (ACM SIGSOFT) *www.acm.org/sigsoft/*
IEEE Technical Council on Software Engineering (TCSE) *www.tcse.org/*
Society for Concurrent Engineering (SOCE) *www.soce.org/*
The International Function Point Users' Group (IFPUG) *www.ifpug.org/*
SPIN–Southern California *www.ics.uci.edu/IRUS/spin/spin.html*
IEEE Computer Society *www.computer.org/*

Professional Networking (private job narket)

American Marketing Association (AMA) *www.svama.org/*
Armed Forces Communications & Electronics Association (AFCEA) *www.afcea.org*
Asian American Manufacturing Association (AAMA) *www.aamasv.com/*
Asian-Silicon Valley Connection (ASVC) *www.asvc.org/*
BayCHI (computer-human interaction) *www.baychi.org/*
Bay Area Council (BAC) *www.bayareacouncil.org/*
Canadian Advanced Technology Alliance *www.cata.ca*
Chinese Information & Networking Association (CINA) *www.cina.org/*
Chinese Software Professionals Association (CSPA) *www.cspa.com/*
Churchill Club *www.churchillclub.com/*
Commonwealth Club *www.commonwealthclub.org/*
Executive Speakers Bureau *www.executivespeakers.com/*
Financial Executives Institute (FEI) *www/feiscv.org*
Forum for Women Entrepreneurs (FEW) *www.few.org/*
Grace Net (Grace Hopper) *www.gracenet.net/*
Hua Yuan Science & Technology Association *www.huayuan.org/*
IEEE–Santa Clara Valley Section *www.ewh.ieee.org/r6/scv*
Japan Society of Northern California *www.usajapan.org/*
Joint Venture Silicon Valley Network (JVSVN) *www.jointventure.org/*
MIT/Stanford Venture Lab *www.venturelab.org/*
Monte Jade *www.montejade.org/*
NASA Ames Research Center *www.arc.nasa.gov/*
Product Management & Development Association (PDMA) *www.pdma.org/*
Professional Business Women of California (PBWC) *www.pbwc.org/*
Public Relations Society of America (PRSA) *www.prsa.org/*
SRI International–Innovation Forum Series *www.sri.com/*

San Francisco Chamber of Commerce *www.sfchamber.com/*

San Jose Business Journal *www.sanjose.bcentral.com/*

San Jose Silicon Valley Chamber of Commerce *www.sjchamber.com/*

Silicon Valley Association of Start-Up Entrepreneurs (SVASE) *http://svase.org/*

Silicon Valley Chinese Wireless Association (SVCWA) *www.svcwireless.org/*

Silicon Valley Fellowship *www.svfellowship.org/*

Silicon Valley Linux Users Group (SVLUG) *www.svlug.org/*

Silicon Valley Manufacturing Group (SVMG) *www.svmg.org/*

Silicon Valley Product Management Association (SVPMA) *http://svpma.tripod.com/*

Silicon Valley Roundtable (SVR) *www.svrt.com/*

Silicon Valley World Internet Center (SVWIC) *www.worldinternetcenter.com/*

Small Business Development Center of Silicon Valley (US SBA) *www.siliconvalley-sbdc.org/*

Society for Competitive Intelligence Professionals (SCIP) *www.scip.org/*

Society of Women Engineers (SWE) *www.swe.org/*

Software Development Forum *www.sdforum.org/*

Stanford University–Breakfast Briefings *http://breakfastbriefings.stanford.edu/*

Stanford University–Technology Ventures Program *www.stanford.edu/group/stvp*

Stanford University–US-Japan Technology Management Center
    http://bases.stanford.edu/USATMC

Straight Talk *www.mppc.org/mppc_ministries/adult_ministries/straight_talk/straight_talk.jsp*

TEC International (International organization of CEOs) *www.teconline.com/*

The IndUS Entrepreneur *www.tie.org/*

Silicon Valley Chapter *www.tiesv.org/*

Toastmasters International *www.toastmasters.org/*

US Chamber of Commerce *www.uschamber.org/default.htm*

Venture Capital Taskforce *www.vctaskforce.com/*

Web Masters Guild *www.webguild.org/*

Women in Technology *www.witi.org/*

Women's Technology Cluster *www.womenstechcluster.org/*

World Affairs Council *www.wacsf.org/*

Xerox PARC–Forum Speaker Series *www.parc.xerox.com/parc-go.html*

Calendars and Industry Events

Broadband Wireless World Forum *www.broadband-wireless.com/*

CeBit *www.cebit.de/*

Comdex *www.comdex.com/*

CTI Expo *www.ctexpo.com/*

eBusiness Expo *www.ebusinessexpo.com/*

Huan Yuan Science & Technology Association–2002 Conference *www.huayuan.org/event.htm*

IEEE–GRID (Calendar of Events) *www.ee.com/grid/grid01.02/sum.htm*

IEEE Communications Society Industry News Cache (email update) *www.comsoc.org/inc*

In and Around the Valley (Churchill Club–e-mail events newsletter)

Chrchll@BenjaminGroup.com Internet World (events calendar) *www.internetworld.com/events*
Linux World *www.linuxworldexpo.com/*
Microsoft–Northern California news and events *www.microsoft.com/usa/norcal*
Networld & Interop *www.interop.com/*
Oracle World Conference *www.oracle.com/start/openworld/oow_sf.html?src=668068&Act=72*
Palm Source (Developers Conference) *www.palmsource.com/*
RSA Security Conference *www.rsasecurity.com/events*
Santa Clara Convention Center *http://santaclara.org/convention.html*
San Francisco Moscone Center *www.moscone.com/*
San Jose Convention Center *www.sjcc.com/*
Silicon Valley Chinese Wireless Association (SVCWA) *www.svcwireless.org/programs.htm*
Silicon Valley Technology and Homeland Security Summit *www.svtechsummit.com/*

Software Development Forum (Events Outlook–e-mail subscription)

events-request@sdforum.org SuperComm (TIA): *www.supercomm2001.com/*
SuperNet–Broadband: *www.supernet2002.com/*
Telecomm–ITU: *www.itu.int/ITUTELECOM*
Wireless Advantage (c/o Symbol Technologies): *www.wirelessadvantage2001.com/*
Wireless CTIA 2001: *www.ctiashow.com/*
Wireless Forum: *www.intmediaevents.com/ewf/spring02/index.html*
Wireless Systems: *www.wirelesssystems2002.com/*
World Economic Forum: *www.weforum.org/*

Vendor Alumni Communities

3Com *www.entre-nouspartners.com/Local%20Settings/Temporary%20Internet%20Files/OLK17/*
     www.corporatealumni.com/after3com
Cisco *http://groups.yahoo.com/group/cisco-alumni*
Hewlett Packard *http://groups.yahoo.com/group/hp_alumni*
Nortel Networks *http://groups.yahoo.com/group/ExNortel*
Oracle *http://groups.yahoo.com/group/oracle-alumni*
Sun Microsystems *http://groups.yahoo.com/group/sun_alumni*

University Alumni Associations

Columbia University *http://columbiaalum.webcountry.net/business.html*
Duke University *www.dukealumni.com/*
George Washington University *www.gwu.edu/alumni.html*
Harvard University–Business School *www.hbsanc.org/*
MIT *www.mitcnc.org/*
Princeton University *www.princeton.edu/Siteware/Alumni.shtml*
Stanford University *www.stanford.edu/home/alumni/index.html*
Yale University *www.yale.edu/alumni*
University of Chicago *www.alumni.uchicago.edu/*
University of Michigan *www.umich.edu/~umalumni*

University of Oxford *www.ox.ac.uk/*

Internet Career Portals and Candidate/Job Seeker Networking (public job market)

6 Figure Jobs *www.6figurejobs.com/*

Bay Area Career Builder *www.bayarea.com/careerbuilder*

Bay Area Careers *www.BayAreaCareers.com/*

Brain Buzz *www.brainbuzz.com/*

Brass Ring *www.brassring.com/*

CGT IT Jobs *http://groups.yahoo.com/group/CGT-IT-Jobs*

C-Six *http://groups.yahoo.com/group/csixnetwork*

Subscribe: csixnetwork-subscribe@yahoogroups.com

Career Action Center (CAC) *www.cac.org/*

Career Journal–Wall Street Journal *www.careerjournal.com/*

Cal Jobs *www.caljobs.ca.gov/*

Computer Jobs *www.computerjobs.com/*

Craig's List *www.craigslist.org/*

DICE *www.dice.com/*

High Technology Careers *www.hightechcareers.com/*

Hire Top Talent *www.hiretoptalent.com/*

Hot Jobs *www.hotjobs.com/*

Job Find *www.jobfind.com/*

Job Hunters Bible (c/o Dick Bolles - *What Color is Your Parachute?*) *www.jobhuntersbible.com/*

Job Smart *http://jobsmart.org/*

Job Smarts for Twenty Somethings *www.jobsmarts.com/*

JobOptions.com *www.joboptions.com/*

Just Tech Jobs (40 technology-specific sites) *www.justtechjobs.com/*

KIT List–Keep in Touch *http://groups.yahoo.com/group/KITlist*

KITlist-subscribe@yahoogroups.com

Layoff Lounge *www.layofflounge.com/*

MPPC Prayer Works *http://groups.yahoo.com/group/MPPC-PrayerWorks*

MSN Careers (see Job Hunt–Who's Hiring) *http://careers.msn.com/*

Marketing Jobs *www.marketingjobs.com/*

Monster.com *www.monster.com/*

Net Temps *www.net-temps.com/*

Opportunity Knocks (Non-profit sector) *www.opportunitynocs.org/*

Pro Match of Silicon Valley (Sunnyvale, CA) *www.promatch.org/*

Professional & Technical Diversity Network (PTDN) *http://yoda.shpe-sv.org/ptdn*

Sales Heads *www.salesheads.com/*

U Post *www.upost.com/*

Vault *www.vault.com/*

WSA Executive Job Search Center *www.wsacorp.com/*

Wednesday Job Group *http://groups.yahoo.com/group/WednesdayJobGroup*

Research and Analysis

Aberdeen Group *www.aberdeen.com/*
Burton Group *www.burtongroup.com/*
Cahners In-stat Group *www.instat.com/*
Dell'Oro Group *www.delloro.com/*
eMarketer *www.emarketer.com/*
Forrester Research *www.forrester.com/*
Frost & Sullivan *www.frost.com/*
Gartner Group *www.gartnergroup.com/*
Giga Information Group *www.gigaweb.com/*
IDG *www.idg.com/*
Infonetics *www.infonetics.com/*
Jupiter Media Mix *www.jmm.com/*
Meta Group *www.metagroup.com/*
Tolly Group *www.tollygroup.com/*
Tom Swift Marketing (San Carlos, Calif.): John Randall, Robert Miller Yankee Group
     www.yankeegroup.com/

High Technology Internet Portals, Standards, and Publications

ATM Forum *www.atmforum.com/*
Aviation Week's Aviation Now *www.aviationnow.com/*
Bandwidth Place *www.bandwidthplace.com/*
Bio.com *www.bio.com/*
Bio Online *www.bio-online.com/*
Byte and Switch–The Storage Networking Site *www.byteandswitch.com/*
Cable Data Communications News *www.cabledatanews.com/*
Channel Web *www.channelweb.com/*
Check Point Software–Security Resources *http://cgi.us.checkpoint.com/rl/resourcelib.asp*
Ciena–Optical University *www.ciena.com/opticalU/home.html*
Cisco–Networking Technology Standards *www.cisco.com/public/products_tech.shtml*
CMP Net *www.cmp.com/*
Comdex (Technical white papers) *http://whitepapers.comdex.com/*
Computer Telephony (Voice & data convergence) *www.computertelephony.com/*
Converge Digest (Voice & data convergence) *www.convergedigest.com/*
CRM Community (customer relationship management) *www.crmcommunity.com/*
CRN–Computer Reseller News *www.crn.com/*
Fiber Optics On-line *www.fiberopticsonline.com/*
Fiber Optics - Industry Associations *www.fpnmag.com/link.htm*
Fiber Optics Product News *www.fpnmag.com/*
Fortune - 100 Fastest Growing US Companies *www.fortune.com/*
Global Information, Inc. (telecomm & IT market portal) *www.gii.co.jp/*
IEEE Communications Society *www.comsoc.org/*
IEEE Standards Association *http://standards.ieee.org/*
Interactive Week *www.interactiveweek.com/*
Internet Telephony *www.itmag.com/*
Internet Week *www.internetweek.com/*

Juniper–Networking Tech Center/Standards *www.juniper.net/techcenter*

Light Reading (fiber optics resource center) *www.lightreading.com/*

Light-Wave (fiber optics) *www.light-wave.com/*

Linux World *www.linuxworld.com/*

MPLS Forum *www.mplsforum.org/*

MPLS Resource Center *www.mplsrc.com/*

MIT Technology Review *www.technologyreview.com/*

Network Computing *www.networkcomputing.com/*

Network Magazine (CMP) *www.networkmagazine.com/*

Network World *www.nwfusion.com/*

Networld & Interop (white papers) *http://whitepapers.interop.com/*

Optical Ethernet Resource Center *www.optical-ethernet.com/optical/index.htm*

Optical Internetworking Forum *www.oiforum.com/*

SANS Institute (security) *www.sans.org/newlook/home.htm*

Silicon India *www.siliconindia.com/*

Silicon Valley Mercury News - Business Section
    www.mercurynews.com/mld/mercurynews/business/

Sunday Spotlight - stock report (hard copy) Sun Guru *www.sunguru.com/*

Tech Target (20 technology-specific portals) *www.techtarget.com/*

Upside *www.crmcommunity.com/*

Wireless Ethernet Compatibility Alliance (WECA) *www.wi-fi.org/*

World of Wireless Communications *www.wow-com.com/*

Standards Bodies, Government Agencies, and Technical Associations

Armed Forces Communications & Electronics Association (AFCEA) *www.afcea.org/*

AFCEA Silicon Valley Chapter *www.geocities.com/sv_afcea/SV_AFCEA.html*

American National Standards Institute (ANSI) *www.ansi.gov/*

Apache Software Foundation (Open source movement) *www.apache.org/*

ATM Forum *www.atmforum.org/*

Biotechnology Industry Organization (BIO) *www.bio.org/*

Broadband Wireless Internet Forum (BWIF) *www.bwif.org/*

CableLabs *www.cablelabs.com/*

CommerceNet *www.commercenet.org/*

EuroCableLabs *www.eurocablelabs.com/*

Federal Communications Commission (FCC) *www.fcc.gov/csb*

Gigabit Ethernet Alliance (GEA) *www10gea.org/*

ICANN *www.icann.org/*

International Biometrics Industry Association (IBIA) *www.ibia.org/*

International Electro-technical Consortium (IEC) *www.iec.com/*

IEEE *www.ieee.org/*

Internet Engineering Taskforce *www.ietf.org/*

International Telecommunications Union (ITU) *www.itu.int/*

Lawrence Livermore Laboratories (US Dept. of Energy) *www.llnl.gov/*

Mobile Applications Initiative (MAI) *www.mobileapplicationsinitiative.com/*

NASA Commercial Technology Network *www.nctn.hq.nasa.gov/*

National Cable & Telecommunications Association (NCTA) *www.ncta.com/*
National Institute of Standards & Technology *www.nist.gov/*
Project Management Institute (PMI) *www.pmi.org/*
Semiconductor Industry Association (SIA) *www.semichips.org/*
Society of Cable Telecommunications Engineers (SCTE) *www.scte.org/*
Storage Networking Industry Association (SNIA) *www.snia.org/*
Telecommunications Industry Association (TIA) *www.tiaonline.org/*
Tele-Management Forum *www.tmforum.org/*
Wall Street Technology Association (WSTA) *www.wsta.org/*
Wireless Communications Alliance (WCA) *www.wca.org/*

Most of the Rest of All the Information You'll Ever Need

Association for Corporate Growth (ACG) *www.acg.org/sv*
Association for Financial Professionals (AFP) *www.afponline.org/*
Classmates.com *www.classmates.com/*
Digital MBAs *www.digitalmbas.org/*
Euro Circle *www.eurocircle.com/*
Garage.com *www.garage.com/*
Job Smarts for Twenty Somethings *www.jobsmarts.com/*
IRS - Itemized deductions (Schedule A - Job Search Expenses) *www.irs.gov/*
Institute for Internal Auditors (IIA) *www.sjiia.org/*
Institute for Management Accountants (IMC) *www.ima-pa.org/*
Keizai Society *www.keizai.org/*
Massachusetts Institute of Technology (MIT) *http://web.mit.edu/*
National Association for Corporate Treasurers (NACT) *www.nact.org/*
National Association of Corporate Directors (NACD) *www.svnacd.org/*
Northern California Human Resources Association (NCHRA) *www.nchra.org/*
One Source *www.onesource.com/*
Progress & Freedom Foundation *www.pff.org/*
San Francisco Chronicle 500 *www.sfgate.com/chronicle500*
Silicon Valley 150 *www.siliconvalley.com/research/sv150*
Stanford University *www.stanford.edu/*
Strategic Capital Concepts, Inc. (investment planning) *www.strategiccos.com/*
The Deal.com *www.thedeal.com/*
US Patent & Trademark Office *www.uspto.gov/*
Moneytree *www.siliconvalley.com/research/vcsurvey*
Venture Reporter.Net *www.venturereporter.net/*
Venture Wire *www.venturewire.com/*
Venture Wire - Executive Summits *http://events.venturewire.com/followon*
Washington Speakers Bureau *www.washspkrs.com/*
Work It - start-up resource portal *www.workit.org/*
Yahoo Groups *http://groups.yahoo.com/*
Yahoo - People & Email Finder *http://people.yahoo.com/*
Young Entrepreneurs Organization (YEO) *www.yeo.org/*
Young Presidents Organization (YPO) *www.ypo.org/*
4Marketeers (roster of trade associations) *http://4marketeers.com/tradeaso.htm*

Index

A

accomplishments, on value-based resume, 49-50
  selecting, 53
accounting job boards, 188
action verbs, 134
  list of, 135
action, ability to take, 95, 96
active interview, the, 94-121
advancement, key to, 20-21
affiliations and associations, on resume, 45
alumni associations, list of, 206
associations,
  list of, 203-210
  using for research, 88
attention, grabbing, 46
awards and patents, on value-based resume, 55

B

behavioral interview, 107-108
  sample questions for a, 182-187
Blur, 50
Bolles, Richard, 14

branding, personal, 50-52
broadcast letters, 59-60
  examples, 61-62
bullets, using on resume, 47

C

calendars and industry events Websites, 205
Career History worksheet, 132
career history, on value-based resume, 54-55
career makeup, 21
chronological resume, 40
  example of a, 41
communication skills, highlighting, 95
competitors, targeting, 87
cover letter as first contact, 93
cover letter rules, 59
covering letters, 60
  example, 64
Covey, Steven, 26

D

direct marketing with e-mail, 77-80
direct response letter, example, 70

About the Author

As a founder of Perry-Martel in 1988, David Perry launched the first information technology search firm, which today dominates the Canadian technology space.

A veteran of more than 800 projects, David is a student of leadership and its effect on organizations, ranging from private equity ventures to global technology corporations.

David has built an extensive personal network of leading CEOs as well as up-and-coming future CEOs, permitting him to find the most qualified candidates to grow a company and increase shareholder value. His **Knowledge-Value** methodology allows clients to rapidly assess critical human qualities, track record, and experience.

David is frequently quoted on trends and issues regarding executive search, recruiting, and HR in leading national business publications including *The Wall Street Journal, IT World, Canadian Business, Venture Wire, Computing Canada, EETimes, NetWork World,* and *HR Today,* and he appears regularly as an executive search and market analyst for *CBC News World.*

David is the Vice-Chair of the *Canadian Technology Human Resources Board* and former board member of the *Software Human Resources Council.* As Business Development Executive of the *Canadian Advanced Technology Alliance (CATA),* David has developed an extensive knowledge of leadership, innovation, and technology. This ever-evolving expertise keeps him at the pulse of most innovative and successful leaders.

David graduated McGill University in 1982 with a Bachelor of Arts in economics. In 1999 he was honored by the *Ottawa Business Journal* as one of the "Top Forty Under Forty" entrepreneurs.

Over the last 18 years, he's staffed hundreds of senior executives in all areas: CEOs, COOs, CFOs, CTOs, CIOs, and other senior technology professionals in sales, marketing, finance, and engineering.